Farley Nobre

Cognitive Machines in Organizations

Farley Nobre

Cognitive Machines in Organizations

Concepts and Implications

VDM Verlag Dr. Müller

Imprint

Bibliographic information by the German National Library: The German National Library lists this publication at the German National Bibliography; detailed bibliographic information is available on the Internet at http://dnb.d-nb.de.

Cover image: www.purestockx.com

Publisher:
VDM Verlag Dr. Müller Aktiengesellschaft & Co. KG , Dudweiler Landstr. 125 a, 66123 Saarbrücken, Germany,
Phone +49 681 9100-698, Fax +49 681 9100-988,
Email: info@vdm-verlag.de

Zugl.: Birmingham-UK, The University of Birmingham, PhD Thesis, 2005

Produced in USA and UK by:
Lightning Source Inc., La Vergne, Tennessee, USA
Lightning Source UK Ltd., Milton Keynes, UK
BookSurge LLC, 5341 Dorchester Road, Suite 16, North Charleston, SC 29418, USA

ISBN: 978-3-639-06862-7

DEDICATION

To Dirceu and Ida (my illustrious parents)
And to all the people who encourage education in the world.

TABLE OF CONTENTS

2

PREFACE

From a macro perspective, this book is about organizations, *cognitive machines* and the environment. It is concerned with the relations between them. It relies on the premise that the technology of *cognitive machines* can improve the cognitive abilities of the organization; and it also relies on the proposition that an increase in organization cognition reduces the relative levels of uncertainty and complexity of the environment with which the organization relates. Moving further, such a proposition opens new perspectives to the study of the implications of *cognitive machines* for organization design, behaviour, cognition, learning and performance.

From a micro perspective, this book is concerned with the design and analysis of *cognitive machines* and it investigates the participation of these machines in dysfunctional conflict resolution which arises from decision-making processes in organizations. It introduces premises and propositions about organization cognition, *cognitive machines* and the participation of these machines in organizations.

It assumes that cognition involves processes which provide individuals, groups and organizations with the ability to learn. It focuses on premises and propositions about organization cognition rather than organizational learning. Therefore, special attention is given to the implications of *cognitive machines* for organization cognition. Its content is chronologically described in the following.

Firstly, it introduces concepts about organization cognition. Such concepts describe relations between organizations, *cognitive machines* and the environment and they are based on concepts of contingency theory, administrative behaviour (decision-making and bounded rationality), open systems and organizational learning. In such a perspective, the organization is viewed as a cognitive system whose cognitive processes are attributes of the participants within the organization and the relationships or social networks which they form. These cognitive processes are supported by the goals, technology and social structure of the organization. Moreover, organization cognition is also influenced by inter-organizational processes and thus by the environment. The participants within the organization comprise humans and *cognitive machines* and they are supposed to act as decision-makers in the name of the organization.

Secondly, it presents a methodology of organization design in order to supports the choice of strategies which increase the degree of cognition of the organization and thus its ability to make decisions. It selects the technology and the participants in the organization as the elements of design since they comprise *cognitive machines*.

Thirdly, it presents the design and analysis of a framework of *cognitive machines*. The design comprises theories of cognition and information-processing systems, and also the mathematical and theoretical background of fuzzy systems, computing with words and computation of perceptions. According to the theory of levels of processing in cognition, it advocates that the ability of these machines to manipulate a percept and natural concepts in the form of words and sentences of natural language provides them with high levels of symbolic processing, and thus with high degrees of cognition. Hence, they mimic (even through simple models) cognitive processes of the human mind. The analysis of these machines comprises theories of bounded rationality, economic decision-making and conflict resolution along with perspectives about their participation in the organization. From the results of the analysis it advocates that such machines can solve or reduce intra-individual and group dysfunctional conflicts which arise from decision-making processes in the organization, and thus they can improve the cognitive abilities of the organization.

Fourthly, it provides evidence by indicating the alignment of its premises and propositions with results of an industrial case study. Its central point of contribution is concerned with the development of approaches and measures to evaluate the degree of organizational cognition. For this purpose it looks carefully at three complementary tasks.

The first task is about processes of organizing. It presents an evolutionary process improvement model - The Capability Maturity Model - which was implemented in the organization of study. This part contributes by defining correlations between measures of organization process improvement and degree of organizational cognition. Among the measures of process improvement are included organization process maturity, capability and performance. From such correlations, it also derives conclusions about the association between organizational cognition and organizational learning.

The second task is about the evaluation of the process of organizing and it proposes the design of a management control system which performs the roles of measurement, analysis and control of the organization process performance. The computation of process performance indexes is performed by a cognitive machine which is engineered with criteria of analysis and design.

The third task is concerned with data analysis, results and conclusions about the industrial case study. In this part, findings indicated that improvements in the level of organization process performance were correlated with improvements in the level of organization process maturity. Such improvements received major contributions from The Capability Maturity Model guidelines. Additionally, improvements in the levels of organization process performance and maturity were associated with improvements in the degree of organizational cognition. These improvements could be measured in two ways. Firstly, on an integer scale [1,5] which indicated the level of organization process maturity associated with the degree of organizational cognition. Secondly, on a real scale [0,10] which indicated the level of organization process performance correlated with the level of organization process maturity, and thus associated with the degree of organizational cognition. A very important implication and contribution of these correlations is that they open new directions to the development of methods to assess, to evaluate and to measure the degree of organizational cognition from appraisal methods of The Capability Maturity Model, and also from other organization process improvement models. It also outlines the main contributions and limitations found with the implementation of The Capability Maturity Model in the organization of study.

Macro results of the industrial case study showed that the Capability Maturity Model (CMM) provided the organization with improvements in its process maturity level. However, despite improving engineering and management processes for complex software projects at the technical level in the organization, the progress of the CMM in those areas of higher hierarchical levels (such as managerial and institutional levels) was slow and poor due to lack of commitment of the organization to the CMM policies at those higher layers.

From such an overview, this book contributes by:

- Playing an important part by choosing selected technologies of machines and connecting them with the disciplines of cognition and organization theory.

- Introducing premises and propositions about organization cognition, *cognitive machines* and the participation of these machines in organizations.

- Presenting a methodology of organization design in order to support the choice of strategies that increase the degree of organization cognition.

- Presenting the design and analysis of *cognitive machines*.
- And providing an industrial case study which comprises practices of organizational cognition along with process improvement.

Dr Farley Simon Nobre – PhD, MSc & BSc
Director CEO - Innovation Technology Enterprise
July 2008 - Brazil

ACKNOWLEDGEMENTS

This book is the largely the result of PhD research undertaken by the author at the University of Birmingham and in the Humboldt University of Berlin between 2001 and 2005, sponsored by CNPq, an Institution of the Ministry of Science and Technology of Brazil. Therefore, I would like to thanks all the people and organizations that contributed to, and participated in the completion of this important project.

I would like to thanks Dr. S.J. Steiner for the reception of me in The University of Birmingham in 2001.

I am very grateful to Dr A.M. Tobias due to his supervision during the writing of my PhD.

I am also indebted to Professor John Child (and his staff) due to his advices given to me during a seminar in the Business School of The University of Birmingham which contributed to improve my knowledge on organizational learning.

My gratitude is also to Professor Kathleen Carley from the Carnegie Mellon University, who introduced me to the field of organization theory by providing me with reference of the most qualified types.

I would like to thanks the staff of The Humboldt University of Berlin due to the services and resources provided by them during my stay in that University in the period between October 2002 and July 2003. In particular, I am indebted to Professor Hans-Dieter Burkhard and Dr. Gabriela Lindemann from the School of Computer Science due to their kind reception and introduction of me to the Artificial Intelligence Research Group, SOCIONICS project, seminars and courses. I also would like to thanks Prof. Dr. D. Demougin, Dr. D. Kübler, and Dr. C. Helm for the lectures provided by themselves on organizations and behavioural economics at the School of Economics and Business Administration of The Humboldt University of Berlin.

I have also greatly appreciated the effort of all professors, scientists and engineers from Brazil who have contributed to the history of this PhD, by providing me with letters of recommendation, by analysing my PhD proposal, by leading me to the background of my B.Sc. degree at PUC-MG (1989-1993) and my M.Sc. degree at UNICAMP (1994-1997), and by sharing with me industrial experiences at NEC (1997-2000).

My scientific acknowledgements are also to both Prof. L.A. Zadeh and Prof. H.A. Simon due to their important contributions to most of the multi-disciplinary fields (spanning from artificial intelligence, cognition and decision analysis to systems theory and organizations) which shaped this research.

Farley Nobre and Prof. L.A. Zadeh in the IFSA World Congress' 1995 (Sao Paulo, Brazil).

PART I: AN INTRODUCTION TO THE BOOK CONTEXT

"The last thing one knows when writing a book is what to put first."

Blaise Pascal (1623-1662).

There is no doubt that the statement above of the French mathematician Blaise Pascal is one of very difficult type and this problem was also found during the writing of Part I.

Part I consists of Chapter 1 which: explains the domain and scope of the book; it overviews the main disciplines which form the body of this work; it describes the problems to be addressed and the solutions to be designed; it provides the motivations for this research; and it explains the whole structure of this investigation. Part I is complemented by Appendixes A, B, C and D.

Appendix A introduces the process of theorizing and the approaches to the study of organizations as adopted in this research; Appendixes B and C surveys the discipline of organization theory; and Appendix D surveys the discipline of technology.

CHAPTER 1. GENERAL INTRODUCTION

1.1. Organizations

1.1.1. The Genesis of Organizations

The practice of organizing is ancient, but formal study of organizations is relatively new. The search for knowledge on organizations through scientific methods of investigation has received increasing attention since the beginning of the 20th century. Such investigations have found enough maturity and formality to constitute a new discipline known today as organization theory.

Principles of organizations evolved with ancient and medieval civilizations, and developed and matured after the Industrial Revolution in Europe in the 18th century and latterly in the United States of America in the 19th century. Such a transformation flourished gradually after the apogee of the Renaissance in Europe which was marked by a period of revolution in thinking, supported by religious, economic, social and political changes [Wren, 1987].

1.1.2. The Role of Organizations of Today

Organizations of today need to be continuously analysed, designed and redesigned on a disciplined and periodical basis. Organization theory constitutes both analysis and design of organizations; it comprises processes of organizing the parts in order to provide efficacy and efficiency[1] to the whole.

Organizations integrate people, technology and goals into a coordinative social structure in order to cope with the environment. The environment includes information; technology; people like consumers and stakeholders; other organizations like buyers and suppliers; and networks of organizations, institutions, market regulators and the whole economy [Milgrom and Roberts, 1992; and Scott, 1998]. The environment also comprises cultural values and natural resources.

Information plays the foremost role in organizations of today since the elements of the organization are contingent upon it. Therefore, organizations have a major task of processing information similarly to a cognitive system with abilities to sensing, filtering and attention, storing and organizing knowledge, problem solving, decision-making and learning [Reed, 1988]. Additionally, in order to pursue intelligent behaviour, organizations have to possess peripheral[2] skills for passing information to the environment. This process of responding and acting complements the previous cognitive stages. Furthermore, information provides the basis for the assertion that organizations shape the environment, and the environment also shapes organizations.

---

[1] Efficacy concerns the attainment of goals and efficiency concerns the economic use of resources in order to satisfy such goals. It does not mean that these concepts are restricted to the perspective of rational systems since natural systems also encompass specific and multiple goals (Chapter 2).

[2] Peripheral is not synonymous with marginal. Instead, it denotes the broader structural features of organizations at the managerial and institutional levels [Scott, 1998].

Studies on organizational learning and knowledge management are presented in [Dierkes *et al*, 2001]; and on organization intelligence, and organizations resembling information processing systems and distributed computational agents are presented in [Blanning and King, 1996; Carley and Gasser, 1999; and Prietula *et al*, 1998]. However, a formal study which relates organizations with concepts of cognition and learning (innovation) was previously proposed in [March and Simon, 1958; and Simon, 1997b]. This book asserts that cognition involves processes that provide individuals, groups and organizations with the ability to learn. Therefore, it focuses on premises and propositions about organization cognition rather than organizational learning.

Tracing back to the industrial revolution, organizations have gradually shifted attention from the conception of corporation of physical structures with agglomeration of people (and machines) to the concepts of processes of organizing[3], information processing systems [Rousseau, 1997; and Sims *et al*, 1993] and learning [Dierkes, 2001]. Therefore, as important as manage themselves, organizations of today have to manage the environment.

1.1.3. The Importance of Organizations

Organizations affect people's lives daily. They provide people with goods, services, well-being, wealth, status, social structures (of normative and behavioural parts), power, etc. Social norms can be conducted by consensus, charisma, legitimacy or force. However, organizations can also provide people with control over others - and thus they can stimulate pathologies [Scott, 1998].

Moreover, people spend much of their time in contact with organizations, as consumers, stakeholders, employees, managers, etc. Hence, people shape organizations. In a broader sense, the economic, social and political facets of local and global cultures shape organizations, and organizations also influence and change them over time.

Organizations exist worldwide in the form of manufacturing and service industries, public and private firms, profit and non-profit institutions, and they include trade and labour unions, schools and universities, armies, churches, hospitals and prisons. IBM, Rolls-Royce, Ford, NEC, British Airways, Banco do Brasil and Lloyds Bank - to name but a few of them - are examples of firms. The formal integration of European markets named European Union is an example of trade union, and thus it is an organization.

1.2. Technology

1.2.1. The Genesis of Technology

Similarly to organizations, technology has provided the history of human evolution and development with major contributions [Gordon, 2000; and Dosi *et al*, 1992]. Technology emerges from the ability of humans to search knowledge, to manage it and to transform it in ends (which can be ends by themselves or other means used to achieve more complex goals). Hence, technology is synonymous with means deliberately employed by humans to attain

---

[3] A corporation is synonymous with an entity organized to pursue goals and to provide goods and services; and the concept of organizing refers to the processes used to make the goods and services possible. Hence, organizing is synonymous with social networks, management and entrepreneurship, cognition and behaviour.

practical outcomes. Therefore, technology encompasses knowledge, and also tools, practices, processes and even other technologies [Kipnis, 1990; and Richter, 1982].

1.2.2. The Role of Technology in Organizations of Today

The domain of technology in organizations encompasses different levels of analysis and distinct elements of application. Analysis ranges from technical, managerial, institutional to worldwide systems, and elements vary from machines to normative and regulative processes.

As stated before, information is the foremost resource used by organizations to cope with the environment, and vice-versa. Hence, technology plays a distinguished role in connecting the flow of information between them. It provides means for the interaction between organizations and the environment, and thus for the exchange of goods, services, people and other resources.

Organizations of today are synonymous with processes of organizing and their functioning resembles a cognitive system with the ability to manage information and to learn. In such a perspective, technology contributes to organizations by providing them with additional means which support them to carry out complex cognitive tasks [Simon, 1977 and 2002].

1.2.3. The Importance of Technology for Organizations

Organizations expand what people can achieve, and technology expands what organizations can do. By integrating these premises, it can be stated that organizations and technology together extend people's capability to achieve more complex goals.

Over the last two centuries, technology has provided organizations with brilliant contributions and with effects on their political, social and economic contexts. Great inventions - to name but a few of them - are those of electricity, including both electric light and electric motors; internal combustion engine, as largely employed in transport systems; petroleum, natural gas, and various chemical, plastics and pharmaceutical processes; communication systems, including the telegraph, telephone, photography, radio and television; urban sanitation infrastructure and indoor plumbing; and medicines, including antibiotics [Gordon, 2000]. Such advances can also include modern wireless communication systems, computers and internet.

Technology infiltrates both organizations and the environment and it provides them with prominent means to cope with each other. Therefore, organizations shape their environment mainly through the use of technology, and the environment also shapes organizations through similar means.

1.3. General Systems Theory

1.3.1. The Genesis of General Systems Theory

General systems theory emerged during the first half of the 20th century paved by the need for a new approach to the unification of sciences, and thus for interdisciplinary concepts, principles, models and laws. It developed as a scientific discipline to deal with principles of organized wholes, including living, nonliving, natural and artificial systems [von Bertalanffy, 1968; and Buckley, 1968]. It received major contributions from the domains of cybernetics [Wiener, 1948, 1954 and 1961], communication theory [Shannon and Weaver, 1963], game theory [von Neumann and Morgenstern, 1944; and Gul, 1997] and systems analysis [Zadeh and Polak, 1969]. Moreover, it got additional insights from principles of complexity [Klir and Folger, 1988; Simon, 1996; and Stacey *et al*, 2000], organization and entropy [Rapoport, 1986], open systems and homeostasis [von Bertalanffy, 1962], self-organizing systems [Ashby, 1968], world and system dynamics [Forrester, 1961 and 1973].

1.3.2. The Role of General Systems Theory

The preponderant characteristic of general systems theory concerns its generality devoted to the study of abstract and mathematical properties of systems, regardless of their physical nature [Zadeh, 1962]. Therefore, the major idea with general systems theorists has been the investigation of isomorphism of concepts, laws and models among various fields spanning from life, earth and natural to social sciences. Hence, useful transfers have been done from one field to another. This has encouraged researchers to develop unified theories [Bahg, 1990; Boulding, 1966 and 1978; Grinker, 1956; and Miller, 1990], and also mathematical and qualitative models for the formulation and derivation of principles which are supposed to be applicable to systems in general [Klir, 1969 and 1972; and Zadeh and Polak, 1969].

Despite such an effort, a gap remains when analysis moves from nonliving to living systems [Buckley, 1968; and Zadeh, 1962]. This gap exists because the laws which govern these two classes of systems are different. Additionally, living systems are generally more complex than nonliving systems. Nevertheless, new approaches have emerged in order to provide alternative tools for the analysis and design of systems of higher levels of complexity [Bond and Gasser, 1988; Weiss, 1999; and Zadeh, 1973, 1997, and 2001].

1.3.3. The Importance of General Systems Theory for Organizations

General systems theorists have provided organizations with new perspectives of analysis and design [Khandwalla, 1977; and Scott, 1998]. Among their main contributions are the concepts of open systems, self-regulation and hierarchical systems.

Open systems constitute a perspective derived from models of biological phenomena [Bertalanffy, 1968]. Put shortly, the open system perspective provides the organization with the definition of a system of interdependent parts interacting with an environment which can be constituted by other organizations or systems [Silverman, 1970]. Through interaction with its environment, the organization can evolve towards an increase of order and complexity [Bertalanffy, 1962]. Such a phenomenon is considered essential to its survival and it provides the organization with the capability of self-maintenance on the basis of a throughput of resources from its environment.

Self-regulation is a principle derived from the broad field of cybernetics [Wiener, 1961], and also from the phenomena of homeostasis within living systems [Bertalanffy, 1962]. Self-regulation means the ability of a system to maintain its steady states by sensing and by responding to its environment. Self-regulating systems encompass processes which work according to some artificial or natural law of behaviour, and they are supported by the principle of feedback. Therefore, they play a fundamental role in control theory, and thus in management control of organizational processes [Anthony, 1984], administrative decision-making [Simon, 1982a], organization design [Haberstroh, 1965] and adaptive learning cycles of learning organizations [Daft and Noe, 2001].

The conception of hierarchic systems is derived from analyses of living systems [Bertalanffy, 1962; and Miller and Miller, 1990], and also from principles of hierarchy of complex systems [Boulding, 1956; and Simon, 1996]. Through this conception, organizations are viewed as systems composed of multiple subsystems, and also systems contained in larger systems called supra-systems. In such a way, organizations can be investigated under different perspectives which span from technical and managerial to institutional levels of analysis [Scott, 1998].

General systems theory influences this research with the domains of unification and isomorphism[4] along with the concepts of open systems, self-regulation, hierarchic systems and complexity.

1.4. The Nature of Theory of Organizations

Part II comprises premises and propositions towards a theory about organization cognition, *cognitive machines* and the participation of *cognitive machines* in organizations. However, in order to propose concepts towards a theory of any subject one first needs to define what one means by theory. Appendix A provides a definition of theory and its components. It introduces a process of theorizing as applied to this book; it presents different approaches to the study of organizations and the criteria to choose research methods among such approaches. Appendix A concludes by explaining the selected approaches to the study of organizations with this research.

1.4.1. The Domain of Theory of Organizations

Theories of organizations fall into the domain of social sciences and they have received most of their contributions from the disciplines of economics, sociology, psychology, politics, management, philosophy and anthropology. Nevertheless, mathematics, engineering and most recently computer science have played prominent tasks in the analysis and design of organizations. They provide organization scientists with tools which support accurate and economic analysis, but also with new theoretical and practical outcomes [Cyert and March, 1963; Gilbert and Troitzsch, 1999; Prietula *et al*, 1998; Simon, 1957 and 1982a; and Starbuck, 1965].

Natural sciences, encompassing physics, biology and chemistry, provides knowledge of theories, principles and laws which are largely derived from the nature through human discoveries. Instead, social sciences derive most of their theories, principles and laws from human and societal behaviour. Therefore, the use of scientific methods for the understanding of people behaviour and cognition constitutes an important part in the social sciences and thus in organizations [March and Simon, 1993].

1.4.2. A Comparative Approach: Theories of Natural vs. Social Sciences

This book assumes that differences between theories of natural and social sciences reside not only in the properties and structure of their elements of study, but most importantly in the abilities of these elements. The former refers to physical, biological and chemical attributes, and the latter means abilities to cognition, intelligence and autonomy.

The main elements of social systems are humans and networks of people, and thus

---

[4] Isomorphism concerns the use of common concepts and approaches to the respective analysis and design of distinct problems and solutions for organizations. Therefore, it is synonymous with economy of resources.

organizations and networks of organizations. Such systems possess high degrees of cognition, intelligence and autonomy which are distributed among their individuals and among their relationships. On the other hand, the elements of, and the relationships with, physical, biological and chemical systems - including all the objects and organisms of the ecological system, but excluding the man - are less complex than those found in social systems since they hold lower degrees of cognition, intelligence and autonomy (if any in many cases).

Therefore, the nature of a theory of organizations resides in principles of human behaviour and cognition.

1.5. A Rationale for New Theories on Organizations

Organization theory has reached the 21st century as a formal and mature discipline, supported by rigorous contributions received from the beginning of the 20th century, and mainly from the last fifty years [Cyert and March, 1963; Dierkes *et al*, 2001; Galbraith, 1977 and 2002; March, 1965; March and Simon, 1993; Milgrom and Roberts, 1992; Pugh, 1997; Scott, 1998; and Simon, 1997b]. The literature about organization theory has provided distinct, complementary and common perspectives of organizations. The publications encompass books which cover different writers of organizations [Pugh *et al*, 1983], diverse types of organizations [McKinlay, 1975; and March, 1965], prominent comparative studies of different classes of organizations [Blau and Scott, 1963], and also references that broadly survey literature results [Hodge, *et al* 2003; and Scott, 1998].

However, organizations and the environment change over time. Not only change their structure and processes of functioning, but also change the perspectives that researchers have about them over periods of time. Hence, one needs to review theories of organizations which relate to the current problem analysis in order to find a better solution design. Therefore, this book comprises the selection of theories of organizations in the literature and also the unification[5] of their concepts towards new perspectives. Special attention is given to the schools of administrative behaviour, decision-making and bounded rationality [March and Simon, 1993; and Simon, 1997a and 1997b], systems theory (open systems) [Buckley, 1968; and Khandwalla, 1977], contingency theory [Galbraith, 1977 and 2002] and also to the perspectives of rational, natural and open systems [Scott, 1998].

1.6. The Scope of the Book

1.6.1. The Domain of Organization Cognition

Firstly, this research focuses on the general picture of organizations pursuing high degrees of cognition in order to reduce the relative levels of uncertainty and complexity of the environment. Therefore, it does not discriminate organizations by their type and purpose (i.e. profit or non-profit industries, public or private institutions, manufacturing and service firms, unions, armies, schools, and so on); nor by their size (number of employees and divisions); nor by their geographical location (east-west); and nor by their age (old-mature or young-immature).

It defines relations between organizations and *cognitive machines*, and also between

---

[5] Unification requires analysis of the various parts of the whole and also design for their integration.

organizations (within *cognitive machines*) and the environment. Such relations are presented in the form of premises and propositions and they include definitions of intelligence, cognition, autonomy and complexity of organizations and machines, along with environmental complexity and uncertainty. Such premises, propositions and definitions form concepts about organization cognition and they support the perspective of organizations as cognitive systems.

1.6.2. The Domain of Organization Design

Secondly, this research presents a methodology of organization design. This comprises strategies to increase the degree of organization cognition and also to reduce the relative level of complexity of the environment. The technology and the participants in the organization are selected as the elements of design because they comprise *cognitive machines*. These machines can be classified as a special type of information technology and they can act in the name of the organization such as decision-makers.

1.6.3. The Domain of *Cognitive Machines*: Analysis and Design

Thirdly, this research focuses on technologies which contribute to model human cognition. Their conception is effective in the sense that they are designed to carry out complex cognitive tasks in organizations.

Proceeding further with the purpose of the organization design methodology, this research presents the design and analysis of a framework of *cognitive machines* [Nobre and Steiner, 2003a].

The design of the framework of *cognitive machines* comprises theories of cognition and information-processing systems [Bernstein, *et al* 1997; Newell and Simon, 1972; and Reed, 1988] and also the mathematical and theoretical background of fuzzy systems [Zadeh, 1965, 1973 and 1988], computing with words [Zadeh, 1996a and 1999] and computation of perceptions [Zadeh, 2001]. Additionally, based on the theory of levels of processing in cognition [Reed, 1988], this book advocates that the ability of these machines to manipulate a percept and concepts in the form of words and sentences of a natural language provide them with high levels of symbolic processing, and thus with high degrees of cognition. Hence, they mimic (even through simple models) cognitive processes of the human mind.

The analysis of these machines involves theories of bounded rationality, economic decision-making [Simon, 1997a] and conflict resolution along with perspectives about their participation in organizations. From the results of the analysis it advocates that such machines can solve or reduce intra-individual and group dysfunctional conflicts which arise from decision-making processes in the organization, and thus they can improve the cognitive abilities of the organization [Nobre and Steiner, 2003b].

1.7. Motivations for this Research

1.7.1. Concepts about Organization Cognition

The perspective of organizations as cognitive systems was put forward by March and Simon [1958 and 1993] and later this was extended by other researchers to the study of organizations as distributed computational agents [Prietula *et al*, 1998]. In such a perspective, individuals have bounded rationality abilities and the organization can extend their limitations to the achievement of more complex goals - e.g. limitations of knowledge, memory and information processing, attention, communication, coordination, decision-making, problem-solving and learning [Carley and Gasser, 1999; and Simon, 1997a and 1997b].

This research supports such a perspective and it focuses on the general picture of the organization pursuing cognition as a necessary feature for its survival, development and achievement of complex goals. Therefore, it introduces concepts about organization cognition which relies on the proposition that an increase in the degree of cognition of the organization reduces the relative levels of uncertainty and complexity of the environment with which the organization relates. What makes this study distinct is the way it connects *cognitive machines* with the organization; and in particular with organization design and with the topic of intra-individual and group dysfunctional conflicts which arise from decision-making processes in organizations.

1.7.2. Organization Cognition and Organizational Learning

Put shortly, organizational learning is a multi-disciplinary field which is concerned with the management and creation of knowledge in organizations [Dierkes *et al*, 2001]. It comprises perspectives of psychology, management, sociology, biology, philosophy, anthropology, history, economics and political science.

This book is about organization cognition and it assumes that cognition comprises processes which provide individuals, groups and thus organizations with the ability to learn, to make decisions and to solve problems. It relates to the field of organizational learning through the perspective of cognitive psychology. In the field of psychology research, cognition is one of the approaches to the study of theories of learning [Lefrançois, 1995]. Therefore, this book contributes to the field of organizational learning by introducing concepts about organization cognition - along with *cognitive machines* and their participation in organizations.

1.7.3. Bringing *Cognitive Machines* closer to Organizations

Machines of today are coming to play an increasing role in organizations as decision-makers. Such machines are emerging to participate and to act in the name of organizations [Nobre and Steiner, 2003a]. Nevertheless, despite being investigated mainly by the literature of engineering and computer science, the subject of machine intelligence and cognition has not received enough emphasis by organization theorists. One of the obvious reasons is that social scientists have little background (and thus knowledge) on the subject of artificial intelligence, computer science, electrical and electronics engineering, among other technological sciences. The other reason is simply the reverse, i.e. technological scientists have little knowledge on organizations and social sciences, and thus they usually apply their engineering, mathematical and computational tools in problems of technical systems.

This research represents an effort to connect the area of organizations with the technology of *cognitive machines* (and vice-versa) in different ways. Firstly, by presenting concepts about organization cognition along with a methodology of organization design which comprises *cognitive machines* as an important element of choice; secondly, by presenting the design of a framework of *cognitive machines* which can reduce intra-individual and group dysfunctional conflicts which arise from decision-making processes in the organization; thirdly, by analysing these machines through the concepts of bounded rationality, economic decision-making and conflict resolution; and lastly by applying a *cognitive machine* in the adaptive learning cycle of an organization whose purpose is the analysis, decision and management control of the performance of large-scale software

26

projects.

This research also attempts to give some steps in order to propose definitions about the roles, responsibilities and relationships between the *cognitive machine*, its designer and the organization.

1.7.4. *Cognitive Machines* and Conflict Resolution in Organizations

Processes of decision-making involve trade-offs among alternatives which are characterized by uncertainty, incomparability and unacceptability, and hence they lead organization's participants to both intra-individual and group dysfunctional conflicts [March and Simon, 1993]. The former conflict arises in an individual mind. The latter type arises from differences among the choices made by two or more individuals and groups in the organization.

Intra-individual and group dysfunctional conflicts also reflect the bounded ability of human sensory organs and brain to resolve details [Zadeh, 2001]. Hence, these conflicts cannot be managed and solved with incentive and reward systems. Such cognitive and information constraints are synonymous with bounded rationality [March, 1994; March and Simon, 1993; and Simon, 1982b, 1997a, and 1997b].

Focusing on this problem, this research contributes by designing a framework of *cognitive machines* which can reduce or solve such conflicts.

1.7.5. *Cognitive Machines*, Bounded Rationality and Economic Decision-Making

The *cognitive machines* as designed in this book are analyzed through theories of bounded rationality and economic decision-making. From the results of the analysis it advocates that such machines can solve or reduce intra-individual and group dysfunctional conflicts which arise from decision-making processes in the organization, and thus they can improve the cognitive abilities of the organization.

Therefore, this research also contributes by putting such machines in the contexts of bounded rationality and economic decision-making along with conflict resolution.

1.7.6. *Cognitive Machines*, Mental Models and Organizations

If one assumes that the cognitive roles in the world of organizations, as fulfilled by agents, have performance and outcomes which can be attributed to either humans or machines, without any distinction, then one is ready to consider machines as members of organizations which act in the name of them similarly to people[6].

This book acknowledges that there is a long way of research before ones have built machines with similar levels of cognition to humans. Nevertheless, it recognizes that important steps have been given with the advancements in computer, artificial intelligence and cognition research from the middle of 20th century.

---

[6] Such an analysis is similar to the classical imitation game of Alan Turing [Turing, 1950], where a machine and a human are placed in rooms apart from a second human being, referred to as the interrogator. Through a terminal of textual communication, the interrogator is asked to distinguish the computer from the human being solely on the basis of their answers to questions asked over this device. If the interrogator cannot distinguish the machine from the human, then, Turing argues, the machine may be assumed to be intelligent.

Proceeding further with such steps, this book presents the design of a framework of *cognitive machines*. Such a design comprises theories of cognition and information-processing systems [Bernstein, *et al* 1997; and Newell and Simon, 1972], and also the mathematical and theoretical background of fuzzy systems [Zadeh, 1965 and 1973], computing with words [Zadeh, 1996a and 1999] and computation of perceptions [Zadeh, 2001]. Additionally, based on the theory of levels of processing in cognition [Reed, 1988], this book advocates that the ability of these machines to manipulate a percept and concepts in the form of words and sentences of a natural language provides them with high levels of symbolic processing, and thus with high degrees of cognition. Hence, they mimic (even through simple models) processes of the human mind.

1.7.7. Application: An Industrial Case

This book provides evidence by presenting an industrial case with NEC. This case is concerned with practices of organizational learning and it comprises two complementary activities: process and technology change management. The first activity is concerned with the implementation of a new process improvement model in one of the industrial plants of NEC. The second activity is concerned with the pilot application of a *cognitive machine* in the adaptive learning cycle of the Radio Engineering Department of NEC do Brasil S.A. It involves tasks of analysis, decision and management control of the performance of large-scale software projects which comprises tangible and intangible outputs. The design of the *cognitive machine* is reinforced with a set of criteria along with qualitative and quantitative analysis. Improvements in the organization process maturity pointed to improvements in both organization cognition and organization performance. Such improvements could be measured in two ways. Firstly, on an integer scale [1,5] which indicates the degree of organization cognition correlated with the level of organization process maturity. Secondly, on a real scale [0,10] which indicates the level of organization performance correlated with the level of organization process maturity.

Therefore, this research also contributes by providing a way to measure the degree of organization cognition by correlating it with the levels of organization process maturity and performance.

1.8. Structure of the Book

This research follows the five stages of the process of theorizing presented in Appendix A. It was designed to include four distinct parts, where each part is associated with one or more stages of the process of theorizing. Its structure is illustrated in Figure 1.1 and described by the following paragraphs.

(Stage 1)   Problem Choice and Analysis

        Part I: An Introduction to the Book Context

(Stage 2)   Solution Design

        Part II: On Organization Cognition and *Cognitive Machines*

(Stages 3 and 4) Data Gathering and Data Analysis

        Part III: Study of Organizations

(Stage 5)   Conclusions

        Part IV: General Conclusions

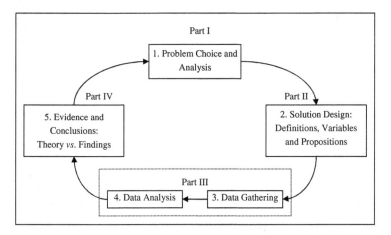

Figure 1.1. Book Structure, Its Parts and Associated Stages

1.8.1. Part I: An Introduction to the Book Context

Part I selects the subject of organizations and the technology of *cognitive machines* as the main areas of research. It consists of Chapter 1 and it is complemented by Appendixes A, B, C and D.

    Chapter 1 briefly overviews the disciplines of organizations, technology and general systems theory. Most importantly, it shows how much these areas of research complement and contribute to each other. This Chapter also explains the nature of theories of organizations; a rationale for new concepts on organizations; the scope of the book which comprises the domains of organization design, the analysis and design of *cognitive machines* and concepts of organization cognition; and it also explain the motivations for this research. Appendix A complements Chapter 1 by introducing a definition of theory and the process of theorizing adopted to this research. It also introduces the criteria used to select approaches to the study of organizations in this book. Appendixes B and C surveys the discipline of organization theory and Appendix D overviews technology.

    Appendix B surveys organization theory. It starts by presenting a perspective on the history of organizations and it continues by describing the roles of organizations of today. It overlooks and reviews the schools of organizations which emerged during the 20th century; and it explains the perspectives of rational, natural and open systems. It introduces rationales for organizing into political, economic and social contexts. It describes the benefits of organizations, the elements which constitute the organization, the concept of organization theory and its relation to the domains of analysis and design. It concludes by explaining the nature of organizations and the sources of diversity of theories of organizations.

Appendix C complements Appendix B by presenting the main disciplines that contributed to the foundation of organization theory.

Appendix D overviews the subject of technology, and it fits technology into the domain of organizations. It starts by presenting a perspective on the genesis of technology and by proposing a definition of technology. It explains the benefits of technology to organizations and it also proposes some rationales for technology into political, economic and social contexts. It explains the scope of technology in organizations, which ranges from different levels of analysis to distinct elements of application. Appendix D continues by explaining why *cognitive machines* can represent a challenge for scientists of organizations and technology. It concludes by analysing the transition of attention from energy to information in the continuous period of Industrial Revolution.

1.8.2. Part II: On Organization Cognition and *Cognitive Machines*

Part II introduces concepts for a theory about organization cognition, *cognitive machines* and the participation of these machines in organizations. It consists of three chapters.

Chapter 2 is concerned with the choice of organization design strategies. It derives premises and propositions about the relations between organization complexity and environmental complexity. While the former is synonymous with organization cognition, the latter is synonymous with environmental uncertainty. It introduces a methodology in order to support the choice of organization design strategies which increase the degree of cognition of the organization and reduce the relative level of complexity and uncertainty of the environment. From such a methodology, the technology and the participants in the organization are chosen as the elements of design because they comprise *cognitive machines*. Chapter 2 also introduces definitions about organizations, the environment and relations between them as viewed throughout this book. Such definitions include concepts for intelligence, cognition, autonomy and complexity of organizations and machines along with environmental complexity.

Chapter 3 is concerned with the design of the organization elements chosen in Chapter 2. Such elements consist of the technology and the participants in the organization and particular attention is given to *cognitive machines*. They were chosen in order to increase the degree of cognition of the organization and thus to improve the ability of the organization to make decisions. Chapter 3 introduces the design of *cognitive machines* with capabilities to carry out complex cognitive tasks in organizations - and in particular the task of decision-making which involves representation and organization of knowledge via concept identification and categorization along with the manipulations of perceptions, concepts and mental models.

Chapter 4 complements Chapter 3 by presenting the analysis of *cognitive machines* through concepts of bounded rationality, economic decision-making and conflict resolution. Such an analysis indicates that these machines can be used to reduce or to solve intra-individual and group dysfunctional conflicts which arise from decision-making processes in organizations. Therefore, they can provide organizations with higher degrees of cognition. Chapter 4 concludes by presenting work relationships between the *cognitive machine*, its designer and the organization.

1.8.3. Part III: Study of Organizations

Part III provides evidence by indicating the alignment of its premises and propositions with results of an industrial case study. Its central point of contribution is concerned with the development of approaches and measures to evaluate the degree of organizational cognition. It consists of Chapter 5 which is complemented by Appendixes E, F and G.

Chapter 5 involves the implementation of a continuous process improvement model in one of the industrial plants of NEC. It also introduces the pilot application of a *cognitive machine* in the adaptive learning cycle of the Radio Engineering Department of NEC do Brasil S.A. It includes tasks of analysis, decision and management control of the performance of five successive large-scale software projects which comprises tangible (e.g. budget) and intangible outputs (e.g. quality and customer satisfaction). The design of the *cognitive machine* is reinforced with a set of criteria along with qualitative (i.e. the phase plane approach) and quantitative (i.e. mathematical and convergence) analysis. Improvements in the organization process maturity pointed to improvements in both organization cognition and organization performance.

Appendix E defines the state variables (X) of the software projects of the Radio Engineering Department of NEC do Brasil S.A.

Appendix F presents linguistic descriptions of the mental models which were designed for the *cognitive machine* in the industrial case.

Appendix G comprises results of the quantitative analysis of the *cognitive machine*.

1.8.4. Part IV: General Conclusions

Part IV provides general conclusions about this research and it consists of Chapter 6 only.

Chapter 6 presents a background about the theories and researchers who most influenced this work. It emphasizes the contributions of this work and it indicates the alignment of its proposal with findings. Topics of further areas of research which can extend this work are pointed out and the chapter concludes by presenting perspectives about the implications of *cognitive machines* for organizations.

PART II: ON ORGANIZATION COGNITION AND *COGNITIVE MACHINES*

"The difficulty lies, not in the new ideas, but in escaping the old ones, which ramify, for those brought up as most of us have been, into every corner of our minds."

John Maynard Keynes (1883-1946).

Part II uses concepts discussed in [Nobre, 2004; and Nobre and Steiner, 2003a and 2003b] and introduces concepts about organization cognition, *cognitive machines* and the participation of these machines in organizations. It relies on the premise and proposition that:

- *Cognitive machines* can improve the cognitive abilities of the organization.

- And an increase in organization cognition reduces the relative levels of uncertainty and complexity of the environment with which the organization relates.

Part II involves Chapters 2, 3 and 4 and its background is largely influenced by:

- The concept of uncertainty, as introduced by the school of contingency theory [Galbraith, 1973, 1977 and 2002].

- Theories of evolutionary and cognitive psychology [Heyes and Huber, 2000; and Simon, 1983]; information-processing systems, perception and cognition [Barsalou, 1999; Bernstein *et al*, 1997; Lefrançois, 1995; Newell and Simon, 1972; and Reed, 1988].

- Theories of fuzzy sets [Zadeh, 1965], fuzzy logic [Zadeh, 1973], computing with words [Zadeh, 1996a and 1999] and computation of perceptions [Zadeh, 2001].

- Theories of cognitive processes in organizations and bounded rationality, as introduced by the school of administrative behaviour and decision-making [March and Simon, 1993; March, 1994; and Simon, 1997a and 1997b].

- The concepts of open systems, complex and hierarchic systems, as introduced by the school of general systems theory [Buckley, 1968; Khandwalla, 1977; and Simon, 1996].

Put shortly, Chapter 2 is concerned with the concept of organization cognition and its relation with *cognitive machines* and the environment.

Chapter 3 introduces the design of *cognitive machines*.

Chapter 4 complements Chapter 3 by presenting the analysis of *cognitive machines* in organizations through the concepts of bounded rationality, economic decision-making and conflict resolution. It concludes by presenting relationships between the *cognitive machine*, its designer and the organization.

CHAPTER 2. ORGANIZATION COGNITION

2.1. Introduction

This chapter relates the discipline of cognition with organization theory and organization cognition with the environment. It relies upon three assumptions:

- Organization complexity is contingent upon organization cognition.
- Environmental complexity is contingent upon environmental uncertainty.
- Organization complexity is contingent upon environmental complexity.

It presents the variables, definitions and propositions which support the concepts about organization cognition. The variables involve cognition, intelligence, autonomy and complexity of both organizations and machines; and also environmental complexity and uncertainty. The definitions comprise perspectives about the organization; the environment; relations between the organization and the environment; and also networks of organizations. The propositions describe relations between organizations, the environment and *cognitive machines*.

2.2. Contingency: Preliminaries on Cognition *vs*. Uncertainty

This section reviews some of the key concepts of contingency theory which play an important part in this chapter.

2.2.1. The Organization and the Environment

Contingency theory has demonstrated through comparative studies and large scale empirical research that organizations are contingent upon the environment. This means that [Galbraith, 1973]:

- "There is no one best way to organize".
- "Not all the ways to organize are equally effective".

    Therefore, organization design is contingent upon the environment

• *The Scope of the Environment*

The environment comprises levels of analysis which can range from technical to institutional aspects [Scott, 1998]. The technical aspect is synonymous with task environment and the organization is viewed as a production system which transforms materials and services from inputs into outputs. In telecommunication companies for instance, there might be divisions for product innovation and research, marketing, systems engineering and design, testing and production. Each division has specific task environments [Nobre and Volpe, 1999 and 2000]. The institutional aspect is broader since it is concerned with the cultural factors along with the belief, normative, regulative and political systems shaping the organization. The perspective of the organization and the environment raging from technical and managerial to institutional and worldwide levels of analysis is presented in the Section 2.6.2.

2.2.2. Uncertainty: Lack of Information and Limit of Cognition

Contingency theory has also defined uncertainty as the variable which makes the organization contingent upon the environment. Hence, organization design, and thus organizational choice, depends on the concept of uncertainty [Galbraith, 1977].

In short, uncertainty is concerned with:

- Lack of information, which leads the organization to unpredictability of outcomes.
- And, insufficiency of cognitive abilities for general information-processing.

The former implies that: "Uncertainty is the difference between the amount of information required to perform a task and the amount of information already possessed by the organization" [Galbraith, 1977].

The latter implies that: Uncertainty is the difference between the degree of cognition required to perform a task and the degree of cognition already possessed by the organization.

These two approaches to uncertainty complement each other since the greater the amount of information which is required to perform a task, the greater the degree of cognition which is required to process this information during task execution.

Figures 2.1 and 2.2 illustrate such concepts of uncertainty using symbolic scales of measurement.

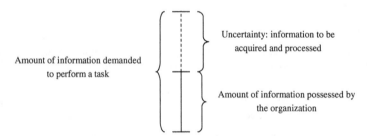

Figure 2.1. Uncertainty as Lack of Information

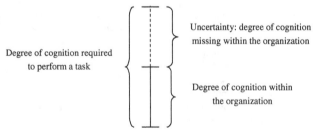

Figure 2.2. Uncertainty as Lack of Cognitive Abilities

Therefore, the question to be answered in the next sections is: - what to do in order to reduce the level of uncertainty with which the organization confronts?

2.3. Organization Design: Strategies and Methodology

Organization design is concerned with the choice of organization models which provide the organization with the ability to cope with the level of uncertainty of the environment with which it relates. It comprises constraint variables and cost-benefit analysis.

Additionally, while the amount of information required from the organization to perform a task depends upon the environment with which the organization relates[7], the degree of cognition of the organization depends upon the choice of its elements.

2.3.1. Strategies for Organization Design

Organization models vary according to the choice of the elements of the organization. The behaviour and the cognitive abilities of the organization are contingent upon the selection of its goals, social structure, participants and technology.

The selection of these elements, and thus organization design, can proceed in two ways:

- Firstly, it can concern the choice of models of organizing which reduce the amount of information that the organization needs to acquire and to process. Therefore, this option reduces the level of environmental uncertainty and keeps unchangeable the degree of cognition of the organization.

- Secondly, it can concern the choice of models of organizing which increase the degree of cognition of the organization. Therefore, this option keeps unchangeable the level of environmental uncertainty and improves the ability of the organization to acquire and to process information.

According to the literature, the first option - i.e. reduction of the amount of information - can be achieved with the strategies for creation of slack resources (i.e. reduction of performance), environmental management (which concerns the organization attempting to influence and to modify the environment, rather than changing its elements) and creation of self-contained tasks (which causes reduction in division of work and thus in the demand for coordination of different tasks). The second option - i.e. the increase of the degree of organization cognition - can be achieved with the strategies for investment in information-processing systems and creation of lateral relations (which concerns with decentralization of decisions) [Galbraith, 1977]. This research supports such strategies and it presents an organization design methodology which includes them. From such a methodology, the strategy for investment in information-processing systems is selected in order to lead the organization to improve its cognitive abilities, and in particular its ability to make decisions.

---

[7] This involves the organization's output diversity (e.g. diversity of goals, products, services, markets, customers, geography, etc.) and level of performance (e.g. strictness of criteria of quality, time and budget constraints, etc.).

2.3.2. A Methodology of Organization Design

The methodology presented here represents a guide to support the decision-maker to select strategies for[8]:

- Reducing the amount of information that the organization must process.
- And, increasing the cognitive abilities of the organization.

This methodology is also used to justify the strategies of organization design selected throughout this research. The strategy selected is concerned with the increase in the degree of cognition of the organization, and hence technology was chosen as the variable of design. The participants in the organization were chosen as an additional variable of design because they comprise *cognitive machines*. This research assumes that such machines can participate and act in the name of organizations such as decision-makers.

The methodology is illustrated in Figure 2.3. The dotted lines symbolize the dynamic interaction between the elements of the organization and a change in one element affects the others. The arrows suggest a path to the process or cycle of design. It starts with the goals and it passes through social structure, technology and participants as a continuous process of analysis, design and redesign.

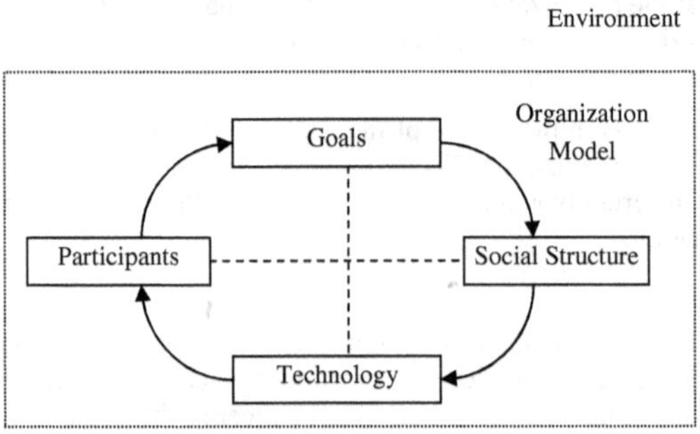

Figure 2.3. A Methodology of Organization Design

- *Goals*

Goals come first because they provide direction to the organization [Galbraith, 2002]. Goals and sub-goals support the units and the participants in the organization in the process of attention[9] [March and Simon, 1993; and Reed, 1988], hence they provide the organization

---

[8] The selection of a strategy involves trade-offs among these options. Moreover, some strategies may cause reverse effects.

[9] Such a process of attention plays the role of directing and focusing certain mental efforts of the organization's participants to enhance perception, performance and mental experience during task execution. This is done when the organization provides specific sub-goals of lateral and vertical relations for its participants and units.

with focalization and reduction of information. Therefore, in this case, reduction of information is a result of goal and sub-goal specification.

The design and redesign of organization goals influence tasks of:

- Creation of Slack Resources: It regards the organization reducing its level of performance [Galbraith, 1977]. Consequently, there is a reduction in information, while the degree of cognition does not change. This is done by relaxing some of the criteria that the organization must attend and also by reducing its output diversity. Creation of slack resources asks for the redesign of the goals of the organization. The extension of the constraint variables of the organization, its divisions and projects – such as time (schedule), budget and effort (participants vs. hours) - demands the organization the use of additional resources and it also may cause lack of credibility with customers. Moreover, the reduction in diversity of goods and services may provoke loss of market and customers. This research assumes this situation is undesirable and unacceptable for organizations of today operating in competitive markets.

- Environmental Management: It regards the organization attempting to influence and to modify the environment, rather than changing its elements [Galbraith, 1977]. Environmental management requires the execution of additional cognitive tasks and it can also ask for the redesign of the goals of the organization. It includes: acquisition and selection of information from the environment (or market); knowledge organization; decision-making and problem-solving. Hence, despite attempting to reduce information, this strategy demands a higher degree of cognition from the organization in order to perform additional tasks and to process additional information. Therefore, it can demand a higher degree of cognition from the organization and thus investments in information-processing systems.

- **Social Structure**

After planning and establishing goals, a social structure is needed in order to support the implementation and the achievement of these goals.

The design and redesign of the social structure can imply for the organization the reduction (or increase) of information and the increase (or reduction) of degree of cognition. Social structure is synonymous with the anatomy and the physiology of the organization. It comprises concepts and design variables such as division of work, departmentalization, span of control and specialization; vertical and lateral processes, rules and decision programs; hierarchy of authority, centralization and decentralization; reward systems; and activities such as the creation of self-contained tasks and lateral relations. These concepts and their implications for the design of organizations are well explained in [Galbraith, 2002; and Scott, 1998] and they have been used by designers in order to find satisfactory[10] results for the organization.

The design and redesign process continues in the following stages with the selection of technology and participants.

---

[10] In this context, which repeats throughout this book, the term "satisfactory" is synonymous with "satisficing" as defined in [Simon, 1997a].

- *Technology*

Information overflows the structure of organizations of today [Galbraith, 2002]. The functioning of the organization depends upon its degree of cognition which provides it with abilities to sensing and perceiving, filtering and attention, storing and organizing knowledge, problem solving, decision-making and learning. Therefore, it is important to choose the technologies which can support the organization with processes and systems that improve its cognitive abilities.

In this research the technology of *cognitive machines* is selected as an element of design in order to improve the degree of cognition of the organization and thus to reduce the level of environmental uncertainty. Particular attention is given to decision-making processes.

- *Participants*

The participants in the organization are the main agents who provide the organization with cognition. Their cognitive, physical, temporal, institutional and spatial limitations are supported by: the social networks which they form; the organization goals and sub-goals which support them in the process of attention; and also the structure of the organization which provides them with managerial and coordination processes.

The participants in the organization pursue cognitive and emotional processes which influence and shape the organization behaviour[11]. Therefore, the selection of participants plays an important part in the survival and development of the organization. However, recruitment and reward systems (with inducements which motivate the participants), and training programs, are processes that the organization must also design in order to succeed.

Therefore, the participants in the organization were chosen as an additional element of design. Particular attention is given to the participation of *cognitive machines* in the organization.

2.4. Variables of the Organization, Machines and the Environment

The concepts derived for the variables in this section represent a synthesis and an extension of some definitions provided within the literature. Cognition and complexity of organizations and machines, and relative complexity of the environment, are the principal concepts to be addressed in this section. The concepts of intelligence and autonomy of both organizations and machines and their relations to cognition and complexity are also introduced.

2.4.1. Organization Intelligence

- *Concepts for Intelligence*

Intelligence is a general mental ability [Schmidt and Hunter, 2000], which depends on one's general cognitive and emotional abilities [Goleman, 1994].

Intelligence depends on two complementary processes which are evoked by abstract or physical stimuli: - they are rational and emotional processes.

---

[11] This book focuses on the cognitive processes of the participants in the organization and it leaves emotional processes for further research.

Rational process or rationality is the ability to follow procedures for decision-making and problem-solving in order to achieve goals [Simon, 1997a]. Rational behaviour is synonymous with intelligence when it leads one to good outcomes [March, 1994]. Hence, the closer the outcome to the optimal solution, the more intelligent is the rational behaviour. Additionally, rational processes are contingent upon the cognitive limitations of humans, and thus they are better represented by the concept of bounded rationality [Simon, 1982b].

Emotional process[12] is less procedural than rationality and it is purposeless in the context of achieving goals. However, researchers have shown that emotions play an important part to motivate, direct and regulate actions in the service of goal pursuit [Bagozzi, 1998; Keltner and Gross, 1999; Keltner, D. and Haidt, H. 1999]. Emotional behaviour is synonymous with intelligence when it represents the ability to excel in life - it includes self-awareness, self-discipline, self-motivation, impulse control, persistence, empathy, zeal, social deftness, trustworthiness and a talent for collaboration. Moreover, on the one hand, emotions influence cognitive tasks such as attention, learning, decision-making and problem-solving [Goleman, 1994]. On the other hand, cognitions are in the service of emotions [Plutchik, 1982] - like in the processes of stimulus interpretation and environment evaluation.

Therefore, intelligence is contingent upon cognitive and emotional processes. Hence, intelligence comprises two complementary elements:

Definition 2.4.1.1: Intelligence

(a) Rational Intelligence: is the ability to use cognitive processes in learning[13], decision-making and problem-solving.

(b) Emotional Intelligence: is the ability to use emotional and cognitive processes in order to understand ourselves (i.e. intra-personal intelligence) and relate with others (inter-personal or social intelligence).

Note: This research focuses more on the subject of cognitive processes rather than emotional processes. Additionally, the concept of bounded rationality is adopted rather than rationality in its classical sense [Simon, 1997a]. Nevertheless, Chapter 3 presents the design of *cognitive machines* with the capability to manipulate a percept and concepts represented in the form of words, propositions and sentences of a natural language. Such concepts can involve emotions[14].

- *Intelligence of Organizations*

Organizations also have intelligence which is provided by their internal elements - i.e. participants, social structure, technology and goals.

---

[12] Emotional process is synonymous with emotion when viewed as a process, rather than simple states [Scherer, K. R. (1982]. One assumes that feeling represents affective and emotional states such as happiness, sadness, anxiety, guilt, fear, jealous, angry, love, etc.

[13] Learning is the process of making changes in the working of our mind, behaviour and understanding through experience [Bernstein *et al*, 1997; and Minsky, 1986].

[14] Linguistic descriptions of emotions include attributes of feelings like happy, sad, angry, etc.

The participants within the organization can provide it with intelligence according to the concepts for intelligence.

The social structure of the organization has normative and behavioural parts. The normative structure provides the organization with rational processes [March and Simon, 1993; and Scott, 1998], and thus with rational intelligence. Complementarily, the behavioural structure can provide the organization with emotional processes [Fineman, 1993], and thus with emotional intelligence.

The technology within the organization can provide it with means which improve its cognitive abilities for learning, decision-making and problem-solving. This comprises information search, attention, representation and organization of knowledge, memory expansion and communication.

Goals and sub-goals provide the organization with focalization and direction. They can support the organization, its units and its participants with the process of attention [March and Simon, 1993]. When viewed as means, sub-goals direct the organization to more complex goals at upper levels. Hence, goals provide the organization with criteria of choice.

2.4.2. Organization Cognition[15]

Organizations resemble cognitive systems when they present abilities and processes for sensing, perceiving, filtering and attention; storing and organizing knowledge; problem solving, decision-making and learning. Such processes are evoked by internal and external stimuli to the organization. Like humans, organizations have relations to the environment.

The perspective of organizations as lateral and vertical distributed cognitive agents was firstly touched upon in the work of March and Simon [1958 and 1993]. Later, it was further extended in the work of Carley and Gasser [1999]. This research adopts the same perspective and it views the structure of the organization resembling a nexus of cognitive agents and processes organized with lateral and vertical relations. Such cognitive agents are the participants within the organization (i.e. humans and *cognitive machines*) and they can also represent a department, a division or a general unit of the organization. They have channels of communication between them which attend the social structure and the protocols of the organization.

In a broad sense, cognition develops in order to increase the probability of humans to survive [Plutchik, 1982]. Similarly, organization cognition plays the same role.

- *Human vs. Organization Cognition*

Human cognition is part of a natural system – i.e. it is not a man-made system – and the brain and thus the cognitive abilities of humans are more a less unchangeable. On the other hand, organization cognition is part of an artificial system[16] – i.e. it is designed and man-made, and

---

[15] Like perception and emotion, cognition is viewed as a process throughout this book. In fact, cognition comprises a set of processes - e.g. attention, knowledge organization, decision-making and problem-solving - which form together the human cognitive system. In such a way, degree of cognition is synonymous with the level of elaboration and integration of such a set of processes.

[16] One also understands that organizations may emerge from informal processes (with no design procedures), but if they want to increase their chance for survival and development, they will need to be reviewed through the process of organization design.

it involves living (e.g. human) and non-living (e.g. machine) forms. Therefore, the cognitive abilities of organizations can be changed and improved through the process of organization design. Organization cognition is contingent upon the goals, social structure, participants and technology of the organization.

- *Organization Cognition vs. Intelligence*

This subsection concludes by presenting premises about organization cognition, degree of organization cognition and by relating organization cognition to intelligence:

Premise 2.4.2.1: Organization cognition comprises a set of processes similarly to those which form the human cognitive system - e.g. processes of attention, knowledge organization and decision-making.

Premise 2.4.2.2: The degree of cognition of the organization is contingent upon the level of elaboration and integration of its associated cognitive processes.

Proposition 2.4.2.1: The greater the degree of cognition of the organization, the greater is its chance to exhibit intelligent behaviour.

2.4.3. Organization Autonomy

Instead of defining autonomy as synonymous with freedom or authority to act, this research regards autonomy as the ability of an organism to act through the use of cognition. Additionally, like cognition, intelligence and complexity, autonomy is a matter of degree. Therefore:

Proposition 2.4.3.1: The greater the degree of cognition of the organization, the greater is its autonomy.

2.4.4. Organization Complexity

This research regards the level of complexity of the organization as contingent upon its degree of cognition. Therefore:

Proposition 2.4.4.1: The greater the degree of cognition of the organization, the greater is its ability to solve complex tasks.

The literature has also defined the level of complexity of an organized social system as a function of the number of components it has, differentiation of its components and interdependence among its components [La Porte, 1975]. Despite this definition does not explicitly concern the complexity of organizations, it only may be related to the social structure of the organization. Furthermore, this definition misses out an important and necessary concept for organizations, the concept of organization cognition.

2.4.5. Machine Intelligence, Cognition, Autonomy and Complexity

The concepts proposed here are derived from the previous definitions of organization intelligence, cognition, autonomy and complexity.

- *Machine Intelligence*

Machine intelligence [Folgel, 2000; and Furukawa *et al*, 1994] and learning [Mitchell, 1997] are all branches of artificial intelligence research [Luger and Stubblefield, 1998]. In short, machine intelligence is a discipline for the design of machines in order to provide them with intelligent behaviour.

The idea of measures of machine intelligence quotient was touched upon in the work of Zadeh [1996b and 1997]. Such an idea is quite reasonable since intelligence is a matter of degree which may be measured in a continuous scale.

However, rather than using the term intelligent machines, this research adopts *cognitive machines*. The reason for such a new nomination is conceptual. Whether machines can have intelligence or intelligent behaviour depends upon how intelligence or intelligent behaviour is defined. Such a definition requires criteria of intelligence which may be accepted or not according to the researcher's perspective and academic background.

Definition 2.4.5.1: *Cognitive machines* are agents whose processes of functioning are mainly inspired by human cognition. Therefore, they have great possibilities to present intelligent behaviour.

- **Machine Cognition**

In short, machine cognition is a discipline for the design of *cognitive machines*. Such machines have a structure which is synonymous with their anatomy, and processes which are synonymous with their physiology and functioning. If the machine structure plays a similar role to the human brain and body (but not necessarily having the same form), the machine processes play a similar role to the human cognitions.

Therefore, the relation between machine cognition and intelligence can be defined by:

Proposition 2.4.5.1: The greater the degree of cognition of the machine, the greater is its chance to exhibit intelligent behaviour.

- **Machine Autonomy**

This research defines machine autonomy similarly to the autonomy of an organism. Hence:

Proposition 2.4.5.2: The greater the degree of cognition of the machine, the greater is its autonomy.

- **Machine Complexity**

This research regards the level of complexity of a machine as contingent upon its degree of cognition. Therefore:

Proposition 2.4.5.3: The greater the degree of cognition of the machine, the greater is its ability to solve complex tasks.

It is important to observe that organization complexity and machine complexity are defined as contingent upon cognition. Therefore, the complexity of organizations and machines are synonymous with their cognitions which are processes used to solve complex tasks.

2.4.6. Environmental Complexity

The complexity of the environment is contingent upon the level of uncertainty that it represents to the organization. Similarly, the complexity of a task environment is contingent upon the level of uncertainty that it represents to the organization during task execution. Hence:

Proposition 2.4.6.1: The greater the level of task complexity, the greater is the level of task uncertainty.

Proposition 2.4.6.2: The greater the level of complexity of the environment, the greater is the uncertainty that the organization confronts.

Note: The level of complexity of the environment is relative to the organization with which it interacts. Therefore, distinct organizations (of different degrees of cognition) confront with different levels of uncertainty even when they operate in a common environment or execute a common task.

2.5. Propositions for the Organization, Machines and the Environment

2.5.1. On Cognition *vs.* Complexity (and Uncertainty)

Firstly, a synresearch of premises is presented to support the propositions introduced in the following.

Premise 2.5.1.1: The elements of the organization - i.e. goals, social structure, participants and technology - support the organization with cognitive processes such as filtering and attention, storing and organizing knowledge, problem solving, decision-making and learning.

Premise 2.5.1.2: The complexity of the organization is contingent upon its degree of cognition.

Premise 2.5.1.3: Technology (and in particular *cognitive machines*) can provide the organization with higher degrees of cognition.

Proposition 2.5.1.1: The technology of *cognitive machines* increases the level of complexity of the organization, and it relatively reduces the level of complexity of the environment.

Firstly, proposition 2.5.1.1 assumes by definition that the higher the level of complexity of the organization, the higher its degrees of cognition. Secondly, it does not mean that the level of complexity of the environment reduces, but that such a level of complexity is relatively reduced when compared to the growth in the level of complexity of the organization. Therefore:

Proposition 2.5.1.2: The higher the level of complexity of the organization, the higher is its degree of cognition.

Proposition 2.5.1.3: The higher the degree of cognition of the organization, the lower is the relative level of complexity of the environment.

Proposition 2.5.1.4: The lower the relative level of complexity of the environment, the lower is the relative level of uncertainty that the organization confronts.

Similarly, proposition 2.5.1.4 says that the level of uncertainty in the environment is relatively reduced with an increase in the degree of cognition of the organization. Therefore, the next theorem can be deduced from the previous chain of propositions:

Theorem 2.5.1.1: The technology of *cognitive machines* increases the level of complexity of the organization (and thus the degree of cognition of the organization), and it relatively reduces the level of environmental complexity (and uncertainty) that the organization confronts.

2.6. Definitions of the Organization

This section proposes definitions of organizations which are complementary to each other.

2.6.1. Organizations as Distributed Cognitive Agents

The definition of organizations introduced in this section represents a synthesis of concepts. The most influential perspectives are those presented in [March and Simon, 1958; Scott, 1998; and Carley and Gasser, 1999].

Firstly, organizations are assemblages of distributed agents. Agents are classified as natural or artificial, and living or nonliving. Humans are natural-living agents, while machines are artificial-nonliving ones.

Secondly, organizations pursue a coordinative system rooted into a social structure which is composed by normative and behavioural parts. Coordinative systems of distinct organizations have different degrees of centralization and decentralization.

Thirdly, organizations pursue goals. The conception of goals varies from individual to organization levels (and also from the technical, managerial and institutional to worldwide levels), and their meaning can range from the perspectives of rational, natural to open systems. Additionally, goal satisfaction is applied rather than goal optimization since agents are given with cognitive, physical, temporal, institutional and spatial limitations (as presented in Appendix B).

Lastly, organizations are open systems, and therefore they pursue the skills of sensing from, and responding to the environment.

In conclusion, organizations are assemblages of distributed and interacting agents with a coordinative system. They are supposed to satisfy[17] goals, and they have relations with the environment.

• *Agents of Organization Cognition*

Cognitive processes are attributes of the participants within the organization and the relationships or social networks which they form. These cognitive processes are supported by the goals, technology and social structure of the organization. Moreover, organization cognition is also influenced by inter-organizational processes and thus by the environment. The participants within the organization comprise humans and *cognitive machines* and they are supposed to act as decision-makers in the name of the organization.

---

[17] In such a context, the term "satisfy" is synonymous with "satisfice" as defined by Simon [1997a].

44

- *Characteristics of the Organization*

(i)   The members of organizations are decision-makers as proposed in [March and Simon, 1993; and Simon 1997b].

(ii) Processes of decision-making involve trade-offs among alternatives which are characterized by uncertainty, incomparability and unacceptability, and hence they can lead organization members to intra-individual conflict. Additionally, members of groups in organizations differ in their perceptions and goals, and thus they can disagree in their decisions causing group conflicts [March and Simon, 1993].

(iii) The intra-individual and group conflicts which arise in organizations as exposed in (ii) are mainly determined by uncertainties and lack of information, and most importantly by cognitive limitations of humans. Hence, these conflicts cannot be solved by incentive and reward systems[18]. Such cognitive and information constraints are synonymous with bounded rationality [March, 1994; March and Simon, 1993; and Simon, 1982b, 1997a, and 1997b]. However, as proposed in Chapter 3, this research also contributes by designing a framework of *cognitive machines* which can be used to reduce or to solve such conflicts.

(iv) The members of organizations have different perceptions. Such a differentiation is accentuated due to the variety of individual motives, but also because of the inequality of distribution of information among the participants in the organization. Therefore, it can lead the participants within the organization to group conflicts [March and Simon, 1993].

(v)   The members of organizations have motives which differ from organization goals. Hence, organizations have to motivate them and to provide them with inducements (such as incentive and reward systems) which lead them to participate in organization activities, including decision-making and problem-solving. If satisfactory alignment is found between the organization and its participants [Gibbons, 1998], then organization equilibrium can be achieved [March and Simon, 1993].

(vi) Organizations shape participants' perceptions and behaviour through social structure, technology and goals, and participants shape organizations through their behaviour, emotions, perceptions, motives and cognitive skills.

(vii) The environment shapes organizations (i.e. their social structure, technology, goals, participants and behaviour), through its sources of complexity and uncertainty, but also through information, services, goods, processes and technology.

(viii) Organizations also shape the environment through similar means.

2.6.2. Organizations as Hierarchic Cognitive Systems

The classification of the organization in technical, managerial and institutional levels of analysis was initially proposed by Talcott Parsons [Parsons, 1960]. This research borrows and supports his ideas and it also extends them to include a fourth level of analysis named worldwide system. Moreover, these levels of analysis are introduced here in the context of cognitive systems. Their meanings are described by the following paragraphs and Figure 2.4

---

[18] A tutorial on strategic reward systems is found in [Dunnette and Hough, 1992: 1009-1055].

illustrates the organization under such a perspective.

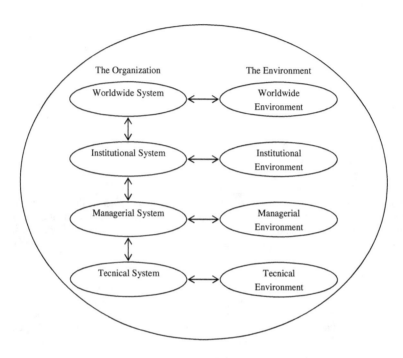

Figure 2.4. The Organization Levels of Analysis

The Technical System: is concerned with cognitive tasks and general activities used for the development of goods and services. It comprises people, machines, communication systems and processes. This level depends on information and resources of the environment for the acquisition of new technologies, and also for the acknowledgement of compliance of goods and services with customers' requirements, technical, quality and general standards.

The Managerial System: is concerned with cognitive tasks of analysis, design and redesign of the organization. It carries out activities of planning, coordination and innovation in areas such as: goals and strategy; structure (normative structure, specialization, span of control, distribution of authority, departmentalization, etc.); technology and processes[19] (of communication, information processing, decision and control); rewards (incentives and

---

[19] Processes can also involve - to name but a few of them - process improvement like CMM [Paulk at al, 1994], quality procedures like ISO 9000 and 14000, principles of management and production like just-in-time and lean-production, intranet and the knowledge to be shared within the organization, policies for recruiting and hiring agents (participants), procedures for evaluating agents and performance, etc.

46

inducements); and human resources (recruiting, training, etc.). Such a level also needs a channel of communication with the environment in order to acquire information about incentive and reward systems provided by other organizations and competitors; to hire new talents; and to select new partnerships with buyers and suppliers for instance. It is also a mediation level between technical and institutional systems.

The Institutional System: is concerned with cognitive tasks used to mediate between the organization and its environment. It comprises the understanding of the social, political, cultural and economic contexts of the organization's environment. The cognitive tasks at this level shape both the technical and the managerial systems, and also the environment (and vice-versa). At this level, participants have responsibility to: understand regulative processes within the market which constrain the boundaries of action of the organization; understand the cultural aspects of the organization and its environment; manage the relationships between the organization and the network of organizations which influence upon its business; understand tax rules on the transaction of goods and services, labour union rights and structure, etc.; set up broader goals and strategies for the organization (like its expansion to other geographical locations and markets, delineation of new products and services, etc.); attract and maintain a body of stakeholders; analyze the wealth of the organization; - to promote partnerships with other organizations; define the percentage of the stocks to be shared within the market; and participate (preponderantly) in the decision processes of design and redesign of the organization.

The Worldwide System: is concerned with cognitive tasks which connect the organization to the world. Such tasks involve the analysis of the implications of organizations, networks and populations of organizations for the social, cultural, economic, political and ecological contexts of the environment. It provides general analysis on the implications of organizations for: the whole economy; the world income distribution; the Gross Domestic Product (GDP) per capita of a country; people social life, well-being, wealth and health; the global ecosystem, its natural resources, energy demand, and so on. Some prominent studies related to this level of analysis are presented in [Easterlin, 2000; Johnson, 2000; Jone, 1997; Pritchett, 1997; and World Bank, 2003].

2.6.3. Organizations as Complex Systems with Cognition, Intelligence and Autonomy

Definition 2.6.3.1: The organization is a special type of dynamic system[20] characterized by a level of complexity C_L which is contingent upon its degree of cognition C_d, intelligence I_d and autonomy A_d.

Axiom 2.6.3.1: Considers that C_L is the level of complexity of an organization O_s and that C_d, I_d and A_d are its degrees of cognition, intelligence and autonomy respectively. Moreover, assumes that C_L can be characterized by a function g of parameters C_d, I_d and A_d:

$$C_L = g(C_d, I_d, A_d) \mid 1 \geq C_L, C_d, I_d, A_d \geq 0 \qquad (2.1)$$

---

[20] A dynamic system has time-varying interactions [Forrester, 1961]. This book views systems as defined in [Bunge, 1987; and Hall and Fagen, 1956]. Additionally, it considers the organization as a system with memory - i.e. given the state of an organization O_s at a discrete time k, then it is assumed that $O_s(k+1) = O_s(k) + O_s(k-1)$.

C_L, C_d, I_d and A_d are defined in the interval [0,1] since they can be characterized by using the concepts of fuzzy sets and membership functions[21] [Zadeh, 1965]. The application of the fuzzy sets theory is encouraged to this definition of organizations because complexity, cognition, intelligence and autonomy are vague and loose concepts in the sense defined by Black [1937 and 1963], and they are also fuzzy in the way defined by Zadeh [1965 and 1973].

<u>Axiom 2.6.3.2:</u> In such a way, let us define an organization O_s denoted here by an object u belonging to a universe of discourse U, which contains the all classes of organizations, i.e. [$u_i \in U \mid i=1,...,N$], for N integer.

<u>Axiom 2.6.3.3:</u> Let us define the level of complexity C_L, and the degrees of cognition C_d, intelligence I_d and autonomy A_d as fuzzy sets with their respective membership functions denoted by $\mu_{CL}(u)$, $\mu_{Cd}(u)$, $\mu_{Id}(u)$ and $\mu_{Ad}(u) \in [0,1]$, i.e.:

$$C_L = \{u \mid \mu_{CL}(u) \in [0,1], u \in U\} \tag{2.2}$$

$$C_d = \{u \mid \mu_{Cd}(u) \in [0,1], u \in U\} \tag{2.3}$$

$$I_d = \{u \mid \mu_{Id}(u) \in [0,1], u \in U\} \tag{2.4}$$

$$A_d = \{u \mid \mu_{Ad}(u) \in [0,1], u \in U\} \tag{2.5}$$

Therefore, O_s can assume four degrees of complexity, intelligence, cognition and autonomy respectively, where such degrees can be interpreted as degrees of compatibility or membership of O_s to the respective fuzzy sets C_L, C_d, I_d and A_d.

From equation (2.1), it can be stated that:

<u>Definition 2.6.3.2:</u> C_L is a function g which can be represented by a t-norm $\cap$ or an s-norm $\perp$ [Dubois and Prade, 1985], i.e.:

$$C_L (\cap) = \{u \mid \mu_{CL}(u) = \mu_{(Cd \cap Id \cap Ad)} \in [0,1], u \in U\} \tag{2.6}$$

$$C_L (\perp) = \{u \mid \mu_{CL}(u) = \mu_{(Cd \perp Id \perp Ad)} \in [0,1], u \in U\} \tag{2.7}$$

---

[21]Fuzzy sets are classes whose boundaries are not clearly defined and hence the transition from membership to non-membership of their elements is gradual rather than abrupt. Examples include the classes of short and tall, young and old, black and white, and poor and rich people. Therefore, the elements v of a fuzzy set A assume degrees of membership $\mu_A(v)$ in A whose values can vary gradually from 0 to 1, in a discrete or continuous way, i.e. $A = \{v \mid \mu_A(v) \in [0,1], v \in V\}$, where V denotes the universe of v. In its broader sense, fuzzy sets theory provides a mathematical background for the representation of information in the approaches to fuzzy logic, computing with words and perceptions [Zadeh, 1973, 1996a, 1999, and 2001].

2.7. Definition of the Environment

This section and the next one are about the environment e, and the relations R_e between the organization O_s and the environment e.

Axiom 2.7.1: Let us consider an organization O_{s1} with relations R_{e1} to an environment e_1 which has relations R_{e2} to another environment e_2. Therefore, a generic environment e_n of an organization O_{sn} may have relations $R_{e(n+1)}$ to another environment $e_{(n+1)}$, where n is an integer.

Axiom 2.7.2: Let us define a network N_E constituted by $(n+1)$ organizations $O_{s(i=1,...,n+1)}$. Let us also define O_{s2} as the environment of O_{s1} with relations R_{e1} between them, and O_{s3} as the one of O_{s2} with R_{e2}. Therefore, it can be derived that $O_{s(n+1)}$ is the environment of O_{sn} with relations R_{en} between them.

Axioms (2.7.1) and (2.7.2) also imply that an environment is relative in the sense that it depends on the position of our analysis on a map of networks of organizations. It also means that the roles of e and O_s may be exchanged since an e becomes an O_s and vice-versa according to the reference of analysis taken on a map of networks of organizations. Therefore:

Definition 2.7.1: Similarly to O_s, definitions (2.6.3.1) and (2.6.3.2) also apply to the environment e (where O_s is replaced with e).

2.8. Definition of Relations: The Organization and the Environment

This section complements the definitions of organizations O_s and the environment e by introducing different types of relations R_e which can exist between them. It borrows the approach to the analysis of ecological dynamics presented in [Boulding, 1978] in order to describe the diversity of R_e.

Axiom 2.8.1: Lets us assume an organization $O_s(t)$ with a set of state variables denoted by $X(t)$, where t denotes time. Additionally, let us define the organization performance $P_{Os(t)}$ as a measure of its efficacy and efficiency[22] which are dependent on the behaviour of $X(t)$.

Axiom 2.8.2: Similarly, let us consider an environment $e(t)$ with state variables $Y(t)$ and with performance denoted by $P_{e(t)}$, which holds the same assumptions given to $P_{Os(t)}$.

Axiom 2.8.3: Let us assume that $O_s(t)$ can affect $e(t)$ in one of three ways. It may affect favourably, and hence the relations $R_e(t)$ is cooperative. A rise in $P_{Os(t)}$ will increase $P_{e(t)}$ (i, if $P_{Os(t)} \uparrow$ then $P_{e(t)} \uparrow$). Secondly, the relationship $R_e(t)$ may be competitive. A rise in P_e leads to a decline in $P_{e(t)}$ and a fall in $P_{Os(t)}$ to a rise in $P_{e(t)}$ (i.e. if $P_{Os(t)} \uparrow$ then $P_{e(t)} \downarrow$ an $P_{Os(t)} \downarrow$ then $P_{e(t)} \uparrow$). Thirdly, $P_{e(t)}$ may have no dependence on $P_{Os(t)}$ and therefore a rise fall in $P_{Os(t)}$ may have no effect on $P_{e(t)}$ (i.e. if either $P_{Os(t)} \updownarrow$ then $P_{e(t)} (0)$).

Similar relations can be postulated for the influence of $e(t)$ on $O_s(t)$. In this case

---

[22] Efficacy concerns the attainment of goals and efficiency concerns the economic use of resources i satisfy such goals.

representations have to be derived. Therefore:

<u>Axiom 2.8.4:</u> Let us denote $R_{(e \to Os)}$ as the relations to the effect of $e(t)$ on $O_s(t)$, and $R_{(Os \to e)}$ of $O_s(t)$ on $e(t)$.

The results of all possible combinations are represented in the Table 2.1 and Table 2.2 describes the results of such combinations.

<u>Definition 2.8.1:</u> Relations R_e are dynamical systems whose attributes can change over time. Examples of attributes applicable to such relations are competition and cooperation. R_e does not guarantee bilateral properties - i.e. the kinds of relations from $O_s(t)$ to $e(t)$ as given by $R_{(Os \to e)}$ may differ from the ones given by $R_{(e \to Os)}$. Moreover, definitions (2.6.3.1) and (2.6.3.2) also apply to the concept of relations R_e between O_s and e.

Table 2.1. Classes of Relationships $R_e(t)$

$R_{(e \to Os)}$ \ $R_{(Os \to e)}$	Cooperative $P_{Os(t)}\uparrow P_{e(t)}\uparrow$	Competitive $P_{Os(t)}\uparrow P_{e(t)}\downarrow$	Independent $P_{Os(t)}\updownarrow P_{e(t)}(0)$
Cooperative $P_{e(t)}\uparrow P_{Os(t)}\uparrow$	1	4	7
Competitive $P_{e(t)}\uparrow P_{Os(t)}\downarrow$	2	5	8
Independent $P_{e(t)}\updownarrow P_{Os(t)}(0)$	3	6	9

Table 2.2. Analysis of Relationships $R_e(t)$

Cases	Interpretation
1	$Os(t)$ contributes to $e(t)$ and $e(t)$ contributes to $Os(t)$
2	$Os(t)$ contributes to $e(t)$ but $e(t)$ harms $Os(t)$
3	$Os(t)$ contributes to $e(t)$ but $e(t)$ has no effect on $Os(t)$
4	$Os(t)$ harms $e(t)$ but $e(t)$ contributes to $Os(t)$
5	$Os(t)$ harms $e(t)$ and $e(t)$ harms $Os(t)$
6	$Os(t)$ harms $e(t)$ but $e(t)$ has no effect on $Os(t)$
7	$Os(t)$ has no effect on $e(t)$ but $e(t)$ contributes to $Os(t)$
8	$Os(t)$ has no effect on $e(t)$ but $e(t)$ harms $Os(t)$
9	$Os(t)$ does not affect $e(t)$ and $e(t)$ does not affect $Os(t)$

2.9. Definition of Networks of Organizations

An important result derived from axiom (2.7.2) and definition (2.7.1) is the concept of networks of organizations as outlined here.

Definition 2.9.1: A network of $(n+1)$ organizations $O_{s(i=1,...,n+1)}$ is a dynamic system denoted by $N_E(t)$ whose relations $R_{e(i=1,...,n+1)}$ change over time.

Relations between organizations and the market change over time. As an example, after the privatisation of the telecommunications market in Brazil in the late of 1990's, most of the companies in that environment lost part of their customers, and since then, they had to find new solutions in order to survive [Nobre and Volpe, 2000].

2.10. Summary

The aim of this Chapter was the establishment of propositions on the relations between organization cognition and environmental complexity.

Organizations are contingent upon the environment which comprises technical and institutional aspects.

Organization design depends on the concept of uncertainty which encompasses lack of information and limit of cognition.

Organization models vary according to the choice of the elements of the organization. The behaviour and the cognitive abilities of the organization are contingent upon the selection of its goals, social structure, participants and technology.

A methodology was introduced to support the choice of organization design strategies which either reduces the amount of information that the organization needs to process, or increases the degree of cognition of the organization. The alternative to increase organization cognition was selected from such a methodology, and technology (of *cognitive machines*) and participants (which include *cognitive machines*) were chosen as the elements of design of the organization.

The concepts of intelligence, cognition, autonomy and complexity of organizations and machines were defined along with the relative complexity of the environment. From such definitions it was established that:

For Organizations:

- Proposition 2.4.2.1: The greater the degree of cognition of the organization, the greater is its chance to present intelligent behaviour.

- Proposition 2.4.3.1: The greater the degree of cognition of the organization, the greater is its autonomy.

- Proposition 2.4.4.1: The greater the degree of cognition of the organization, the greater is its ability to solve complex tasks.

For Cognitive Machines:

- Proposition 2.4.5.1: The greater the degree of cognition of the machine, the greater is its chance to present intelligent behaviour.

- Proposition 2.4.5.2: The greater the degree of cognition of the machine, the greater is its autonomy.

- Proposition 2.4.5.3: The greater the degree of cognition of the machine, the greater is its ability to solve complex tasks.

For the Environment:

- Proposition 2.4.6.1: The greater the level of environmental complexity, the greater is the level of environmental uncertainty.

As for the Environment and the Organization:

- Proposition 2.4.6.2: The greater the level of complexity of the environment, the greater is the uncertainty that the organization confronts.

And as for the Organization, *Cognitive Machines* and the Environment:

- Theorem 2.5.1.1: The technology of *cognitive machines* increases the level of complexity of the organization (and thus the degree of cognition of the organization), and it relatively reduces the level of environmental complexity (and uncertainty) that the organization confronts.

This chapter also introduced the characteristics of organizations as viewed throughout this work. It defined organizations as distributed cognitive agents, hierarchic cognitive systems and complex systems. From such a background it was established that the level of complexity of the organization is contingent upon its degrees of cognition, intelligence and autonomy. It concluded by extending such definitions of the organization to the environment; by presenting definitions of the relations between the organization and the environment; and by defining networks of organizations.

CHAPTER 3. DESIGN OF *COGNITIVE MACHINES*

3.1. Introduction

This chapter introduces the design of *cognitive machines*. It contributes by bringing selected technologies of machines (and artificial intelligence) closer to the discipline of cognition.

In short, Chapter 3 is concerned with the design of the organization elements selected in Chapter 2. Such elements are the technology and the participants in the organization. Particular attention is given to *cognitive machines*. They were chosen in order to increase the degree of cognition of the organization and thus to increase the ability of the organization to make decisions.

Concerning its contents, Chapter 3 presents the design of *cognitive machines* with capabilities to carry out complex cognitive tasks in organizations - and in particular the task of decision-making which involves representation and organization of knowledge via concept identification and categorization along with manipulation of percepts and concepts. The ability of these machines to manipulate a percept provides them with higher levels of information-processing than other symbolic-processing machines[23]; and according to the theory of levels of processing in cognition [Reed, 1988], these machines can mimic (even through simple models) cognitive processes of humans [Nobre and Steiner, 2003a]. Percepts and thus concepts[24] (along with mental models) are described by words, propositions and sentences of natural language [Zadeh, 2001].

3.2. Humans, Cognition and Machines

3.2.1. Evolution of Cognition

Organisms of the ecological system have evolved towards the improvement of their abilities and mechanisms for fitness and adaptation in the environment. Among such organisms, human beings are the species that has found the highest probability to survive, to reproduce and to continue evolving and developing.

Such a predominance of humans is a particular privilege provided by the evolution of their brain, emotional and cognitive processes [Heyes and Huber, 2000; and Simon, 1983]. Among the results of such a continuous evolutionary path are their abilities to learn, to search

---

[23] *Cognitive machines* manipulate complex symbols in the form of words, propositions and sentences of natural language which are descriptions of percepts and concepts. Such complex symbols are codified through the principles of linguistic variables and fuzzy granulation, fuzzy sets and membership functions, and fuzzy generalized constraints; and they are manipulated through the mechanisms of fuzzy logic and fuzzy constraint propagation [Zadeh, 1973, 1975, 1976, 1988, 1996a and 1999]. Therefore, such *machines* can manipulate more complex symbols than other machines whose base of computation is the classical set theory [Halmos, 1960] and crisp granulation.

[24] Put shortly, percept and concepts are alike when a percept is recognized and classified into a category. For example, by saying that Brazil is *large*, ones is assuming that the size of the Brazilian land is classified as *large*, and in fact, *large* is a concept. For simplicity, concepts and concepts are treated as synonymous throughout this chapter.

53

information and to organize knowledge, to make decisions and to solve problems. Humans adapt to the environment, but they also change the environment to their own needs. Humans have cultivated agriculture, improved their immunology system, modified natural resources in order to explore energy, created organizations and cities, and developed transport and telecommunications systems, designed machines which mimic their own behaviour, and so on. In such a way, humans have been transferring some of their abilities to systems, and in particular to machines.

3.2.2. Designing Cognition: From Humans to Machines

Perhaps, cognition is the most precious and difficult ability that humans can transfer to artificial (man-made) systems. The design of cognitive processes requires from humans the understanding of their own mind and the implementation of these processes into machines depends on the availability and development of appropriate technologies. This work advocates that scientists still have a long way of research in order to engineer *cognitive machines* which mimic a complex model of the human mind, but it also considers that they have had enough advancement in such a field in order to form a body of knowledge on the engineering of machines with some cognitive abilities [Haikonen, 2003; Luger, and Stubblefield, 1998; Newell, 1990; and Zadeh, 2001]. Moreover, this work assumes that the birth of the discipline of artificial intelligence in the 1950's was the mark for the beginning of what is called the design of *cognitive machines*.

However, the terms artificial intelligence and machine intelligence are not used in this research for one main reason:

Note: Intelligence and intelligent behaviour depends on cognition and emotion[25]. Therefore, and firstly, one should concern the design of cognitive processes in order to provide machines with some intelligent behaviour. Machines with cognitive processes have greater probability to behave intelligently than any other machine.

Figures 3.1 and 3.2 illustrate two different strategies to the design of machines with intelligent behaviour and cognition respectively. The strategy illustrated in the Figure 3.2 is the one adopted in this research.

---

[25] Intelligence, cognition and emotion are defined in Chapter 2.

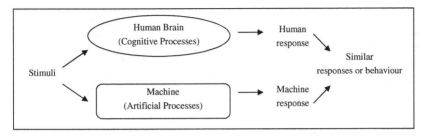

Figure 3.1. Design Strategy with focus on Machine Intelligence and Intelligent Behaviour

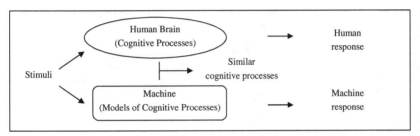

Figure 3.2. Design Strategy with focus on Machine Cognition

3.3. Theories and Technologies behind *Cognitive Machines*

The design of *cognitive machines* comprises theories and models of cognition along with technologies to engineer them. This section overviews the theories and technologies which support the model of *cognitive machines* as concerned throughout this research.

3.3.1. Information Processing Theory: A Cognitive Approach

The literature about cognitive psychology has presented different and unified theories on cognition [Newell, 1990], and also theories on the relation between cognition and perception [Barsalou, 1999]. Information-processing theory is the approach selected to the study and design of cognitive processes in this research. This approach uses the computer as metaphor to simulate and to understand human thinking and processes such as decision-making and problem-solving [Newell and Simon, 1972; Reed, 1988; and Reisberg, 1997]. Therefore, this research uses such an approach as a framework to the design of *cognitive machines*.

3.3.2. The Scope of Design is Decision-Making

According to the literature, processes of cognition include sensing and perceiving, pattern recognition, attention, memory, concept formation and attainment, categorization, verbal and spatial knowledge, representation and organization of knowledge, language, though, comprehension, problem-solving and decision-making [Reed, 1988; and Lefrançois, 1995].

However, this chapter focuses on decision-making and its associated cognitive processes. It concentrates efforts in the design of a framework of machines with the ability to make decisions. According to Chapter 4, these machines are supposed to participate of conflict resolution in organizations. They are likely to contribute with decision-making processes which can reduce or solve intra-individual and group dysfunctional conflicts in organizations. Therefore, this chapter touches cognitive processes of:

- Perception, attention and concept identification;

- Short-term and long-term memory;

- Representation and organization of knowledge via categorization;

- And decision-making.

3.3.3. On Machine Learning and Problem-Solving

Processes of learning and problem-solving are left for further research since the main concern of this research is decision-making. However, the disciplines of neural computation [Hertz, 1991], soft computing [Zadeh, 1994], adaptive fuzzy systems [Wang, 1994], evolutionary computation and genetic algorithms [Back, *et al*, 2000; and Fogel, 2000], along with genetic programming [Koza, J.R. 1992] have demonstrated in the literature to be powerful tools which can aggregate abilities of learning and problem-solving to the *cognitive machine* framework as designed in this chapter.

3.3.4. Machines are Amodal-Symbolic-Processing Systems

Theories of cognition and perception can be classified into modal and amodal-symbol systems [Barsalou, 1999].

- *Modal-Symbol Systems*

In modal-symbol systems, the perceptual states which arise in sensory-motor systems are extracted and selected via the process of attention, and later stored in memory[26] to function as symbols. The structure of these symbols is analogically related to the perceptual states which produced them. On this view, perception and cognition are interdependent processes and they share common parts.

- *Amodal-Symbol Systems*

In amodal-symbol systems, perceptual states are translated or codified into a new representational system of symbols. As a consequence, the internal structure of these symbols is unrelated or only linked arbitrarily to the perceptual states that produced them.

The *cognitive machines* as considered in this research (like computers) operate as amodal-symbol systems. They require artificial transducers to map perceptual states (e.g. the

---

[26] Memory stores are classified in sensory store (SS), short-term memory (STM) and long-term memory (LTM). SS provides a brief storage for information in its original sensory form and it extends the amount of time that a person has to recognize a pattern. STM is limited in both the amount of information it can hold (capacity) and the length of time it can hold the information (duration). LTM has neither of the two limitations of STM. STM holds a relation to LTM since it combines information that is retrieved from LTM with information that arrives from the environment [Reed, 1998].

temperature of a room) to a new base of symbols (which may be represented by numbers like 25°C, words like *warm* and sentences of natural language like *warm but not too warm*).

Figures 3.3 and 3.4 illustrate how modal and amodal-symbol systems function when perceptual states about a car arise in sensory-motor systems. While the perceptual states $(s_{i=1,...,M})$ in the modal-symbol system are mapped to perceptual symbols $(p_{i=1,...,M})$ of analogue internal structure, these same states $(s_{i=1,...,M})$ are mapped to, and thus translated into, new representational structures $(y_{i=1,...,N})$ of functional symbolic systems [Newell and Simon, 1972; and Minsky, 1986]. The output arrows at the right side of both figures point out to the activation of new cognitive functions such as memory and decision-making.

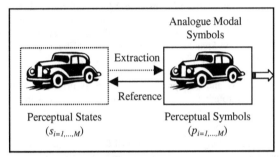

Figure 3.3. Modal-Symbol Systems

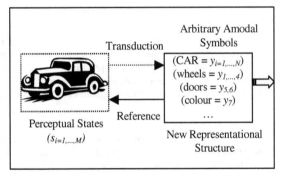

Figure 3.4. Amodal-Symbol Systems

3.3.5. On the Scope of the *Cognitive Machine* Technology

In short, this research considers those machines which operate based on, but are not limited to, one or more principles among electrical, mechanical, analogue, digital, optical, hybrid and artificial cognitive-neural signals. Secondly, such machines are made of technologies of computers, communication networks and software programs. Thirdly and most importantly, this research assumes that if these machines are to pursue high degrees of cognition, then they should be governed by the disciplines of fuzzy systems [Klir and Yuan, 1996; Zadeh, 1973;

Klir and Folger, 1992; and Wang, 1994], computing with words and computation of perceptions [Zadeh, 1996a, 1999, and 2001] along with soft computing [Zadeh, 1994 and 1997]. While the former technologies are synonymous with the anatomy or structure of the machines, the latter technologies (i.e. the disciplines) are synonymous with their physiology, functioning and cognitive processes.

Principal attention is given to the disciplines of fuzzy systems (FS), computing with words (CW) and computation theory of perceptions (CTP). The criteria used to select these disciplines are based on the following pillars:

- Firstly, humans have a distinguished ability to perform diversified physical and mental tasks without any manipulation of measurements [Zadeh, 2001] - such as driving in city traffic and summarizing a speech for instance. When performing such tasks, humans use their ability to perceive objects and sounds, to form concepts and to manipulate them. Most of the concepts humans form and reason with have fuzzy boundaries [Bernstein, *et al* 1997; Lefrancois, 1995; and Reed 1988] - e.g. the attributes of colour like *green* and *blue*; the attributes of price like *cheap* and *expensive*; etc. These concepts are often described by words, propositions and sentences of natural language. In such a way, the disciplines of FS, CW and CTP provide the necessary principles to represent percepts and thus concepts through complex symbols in the form of words, propositions and sentences of natural language. Additionally, they also provide mechanisms to manipulate such symbols.

- Secondly, decision-making and general tasks in organizations involve not only numerical information, but also perceptions and emotions, and thus the identification and manipulation of concepts which have fuzzy boundaries – examples of such tasks include the design of organizations, management and recruitment of people, and most of the activities within the technical, managerial, institutional and worldwide levels of analysis of the organization.

- Thirdly, the theories and mechanisms of computing with words and computation of perceptions provide a background to the design of machines with the ability to process more complex symbols than other approaches [Zadeh, 1999 and 2001]. Therefore, according to the theory of levels of processing in cognition [Red, 1988], such machines can operate at high levels of symbolic processing, and thus they can find high degrees of cognition.

- Fourthly, regarding the discipline of fuzzy systems, it emerged as a new approach to the analysis of complex and decision-making processes such as those found in social systems; and also to provide a bridge between the analysis of man-made (like machines) and living systems (like humans) [Zadeh, 1962, 1965, and 1973].

- Fifthly, they are methodologies which complement and extend the approaches to crisp computation to more complex applications where the available information is too imprecise to justify the use of numbers. Moreover, such methodologies are necessary when there is a tolerance for imprecision which can be exploited to achieve tractability, robustness, low solution cost and better rapport with reality [Zadeh, 1999].

- Sixthly, they were developed for the analysis and design of systems which pursue high degrees of machine intelligence quotient [Zadeh, 1996b and 1997].

- Lastly, such disciplines have found maturation supported by theories and applications [Pedrycz and Gomide, 1998].

3.4. Anatomy of *Cognitive Machines*

3.4.1. A General Structure of Information-Processing Machines

An outline of the *cognitive machine* structure is sketched in the Figure 3.5. This structure is adapted from the information-processing system approach presented in [Bernstein, *et al* 1997; and Newell and Simon, 1972]. In short, such machines operate like:

(1)   Stimuli from the environment are modified and transformed by the sensory system into neural activity signals. These signals are called sensations.

(2)   The perceptual system maps such sensations into new structures and representations of perceptual amodal-symbols. Viewed as a process, perception organizes sensations into patterns, and furthermore, it uses knowledge stored in memory to recognize those patterns. It gives meaning to sensations through perceptions of depth, distance, motion, light, etc.

(3)   The processor receives and manipulates perceptual amodal-symbols. It consists of a process of reasoning which uses the knowledge stored in memory to make decisions. Its output decisions are represented by the same structure of amodal-symbols given by the perception block.

(4)   The memory stores knowledge in the form of mental models described by concepts, categories and clusters of propositions.

(5)   The response block transforms amodal-symbols to a new structure compatible with the environment requirements. This block can also include task execution and actuation on the environment.

(6)   The process of attention acts on the perception, decision-making and response processes. Attention provides perception with the selection of specific parts of stimuli and sensations when recognizing patterns and storing them into memory for further manipulation in decision-making and response.

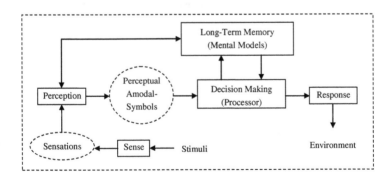

Figure 3.5. Structure of the *Cognitive Machine* as an Information-Processing System

59

3.4.2. A Framework of *Cognitive Machines*

This subsection introduces a framework of *cognitive machines*, and most importantly, it contributes by relating the functioning of such a framework with the processes of cognition.

This framework is tailored from the general structure of the information-processing system illustrated in the Figure 3.5. Its processes and functioning are designed according to the principles of fuzzy logic and fuzzy control [Lee, 1990; Mamdani, 1974; Nobre, 1997; Wang, 1994; and Zadeh, 1968 and 1973], computing with words [Zadeh, 1996a] and computation of perceptions [Zadeh, 1999 and 2001]. Such a framework involves the processes listed in the Figure 3.6 and has its architecture sketched in the Figure 3.7.

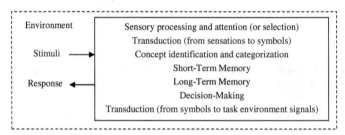

Figure 3.6. Processes associated with the *Cognitive Machine*

Note: The functioning of the framework sketched in the Figure 3.7 is similar to the general structure of the information-processing system presented in the Figure 3.5. However, this framework manipulates percepts and concepts in the form of complex symbols described by words, propositions and sentences of natural language. Most importantly, such a framework is equipped with the machineries of fuzzy logic, computing with words and computation of perceptions [Zadeh, 1965, 1973, 1999 and 2001] in order to manipulate a percept and concepts, clusters of propositions and thus representations of mental models.

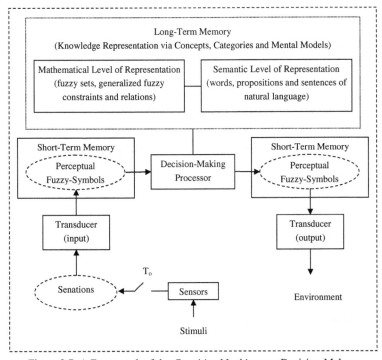

Figure 3.7. A Framework of the *Cognitive Machine* as a Decision-Maker

3.4.3. Levels of Symbolic-Processing

The levels of information-processing of the *cognitive machine* represent layers of a simplified model of the human mind. From a bottom up perspective, the *cognitive machine* maps information from the levels of stimuli and sensations (neural activity signals) to the levels of percepts[27] and thus concepts[28] stored in memory in the form of words, propositions and sentences of natural language. At its higher level of processing, the machine manipulates percepts and thus concepts in the form of clusters of propositions which represent one's understanding of how things work. Such clusters are called metal models [Bernstein, *et al* 1997].

---

[27] Perception is the process through which sensations are interpreted; using knowledge and understanding of the environment, so that they become meaningful experiences [Bernstein, *et al* 1997].

[28] Concepts are categories of physical and abstract objects with common properties like the attributes of colour (*red, yellow, green,* etc), size (*small, medium, large,* etc), etc. A concept may be regarded as a percept recognized and classified into a category.

61

In such a way, fuzzy sets and fuzzy logic along with computing with words and computation of perceptions appear as appropriate tools to represent descriptions of mental models; and secondly, they also provide the necessary mechanisms to manipulate such mental representations similarly to the ability of humans to think with fuzzy concepts along with approximate reasoning [Gupta and Sanchez, 1982; and Sanchez and Zadeh, 1987].

Table 3.1 resumes such levels of processing with the necessary tools for their engineering.

Table 3.1. Levels of Symbolic-Information-Processing

Information level	Processing level	Representation	Technology (Tools)
Stimuli and sensations	Sensory and neural circuit systems	Signals (electrical, optical, digital, etc)	Sensors
Sensations and percepts	Transducer and decision-making processor	Fuzzy-perceptual symbols (linguistic variables)	Fuzzy sets and fuzzy granulation
Concepts, categories and mental models	Memory and decision-making processor	Words, propositions and clusters (natural language)	Fuzzy constraints and modelling

3.5. Processes within the *Cognitive Machine* Framework

This section describes the functioning of the framework presented in the Figure 3.7 by associating its functional blocks to the cognitive processes of Figure 3.6.

3.5.1. Sensory Processing and Attention[29]

The sensor selects stimuli, interprets them and transforms them to sensations. It involves a simplified process of attention.

In short, attention is a process that provides humans with the ability to focalization and concentration on specific information [Reed, 1988]. It directs our sensory and perceptual systems toward certain stimuli and sensations for further processing and storage in memory [Bernstein, *et al* 1997]. In organizations, attention plays the role of directing certain mental efforts of the organization participants to enhance perception, performance and mental experience during task execution.

In the Figure 3.7, a simplified process of attention is implemented by the sensory block. This block selects only part of the information which arrives from a stimulus. The selected and the rejected pieces of information by the sensor are respective synonymous with the definitions of figure and ground in perception [Bernstein *et al*, 1997]. Sensors can operate with analogue or digital principles and they can be classified according to their capability to mimic vision (e.g. light and colour), hearing or audition (e.g. sound), touch (e.g. pressure,

---

[29] Attention encompasses bottleneck and capacity theories. The first theory views attention as a filter that selects the information (stimuli and sensations) to be perceived and recognized (as patterns) and stored in memory (as concepts). The second theory views attention as synonymous with cognitive limitation and it emphasizes the amount of mental effort that is required to a perform task [Reed, 1988].

temperature and pain), olfaction (smell), gestation (taste) and even proprioception.

The key T_0 samples pieces of information which arrive from the sensory block at a discrete period of time called sampling time (KT_0). This key holds the information for a period of time represented by KT_0 whose value is specified according to the task that the *cognitive machine* executes and the environment where it operates. In such a way the key T_0 can be viewed as a sensory store.

3.5.2. Transduction: From Sensations to Fuzzy-Perceptual Symbols

The (input) transducer block is equivalent to a fuzzifier [Jager, 1995; Lee, 1990; Nobre, 1997; and Wang, 1994]. It maps sensations provided by the sensory block (and sampled by the key T_0) to amodal-symbol representations. In other words, the transducer transforms signals at lower levels of meaning to more complex symbols of higher levels of meaning. An analogy can be done for instance by transforming analogue signals, digital codes or numbers like 30°C to a more complex base of symbols in the form of words like *warm* and sentences of natural language such as *warm but not too warm*.

According to the cognitive theory of levels of symbols and processing [Reed, 1998], the human brain carries information in the form of signals at low levels within neural processing circuits. In such a level of processing, neural signals and processes are beyond human power of introspection [Haikonen, 2003]. However, the human brain has a remarkable ability to represent such low level neural signals by more complex symbols and thus mental models. This ability provides humans with higher levels of information-processing and hence they can manipulate information in the form of words, propositions, sentences of natural language and images. Examples of tasks which require high levels of symbolic-processing include driving in city traffic, playing football and golf, cooking a meal, summarizing a story, recruiting and managing people, etc.

For *cognitive machines*, sensations can be presented in the form of electrical, optical and other types of signals. Such signals are mapped by the transducer to fuzzy-perceptual symbols. These symbols are representations of words which are labels of percepts and thus concepts. This mechanism of representation of information through complex symbols provides the *cognitive machine* with high levels of information-processing similarly to simplified models of the human mind.

3.5.3. Fuzzy-Perceptual Symbols

The structure of fuzzy-perceptual symbols is constituted by mathematical representations of percepts and thus concepts. From a top-down approach, fuzzy-perceptual symbols involve three levels of processing. They are illustrated in the Figure 3.8 and described in the following through a top-down perspective.

(a)   The first level is called perceptual and conceptual level of processing. At this level, the symbols which the *cognitive machine* manipulates represent percepts and concepts. Percepts and concepts are alike when a percept is recognized and classified into a category. Concepts are attributes of physical and abstract objects like colour (*blue* and *red*, etc), depth and distance (*short, long*, etc), size (*small, big*, etc), age (*young, medium age, old*), form and shape (*oval, round*, etc), motion and speed (*slow, fast*, etc), price (*expensive, cheap*, etc), *truth, justice, likelihood*, and so on. Such concepts are identified and formed from sensations extracted in one's sensory organs.

(b)   The second level is called natural language level of processing. At this level, concepts and thus percepts are labeled by words, propositions and sentences of a natural language such as: - Mary is *young*; temperature is *hot*; Rob lives *near* to London; the car is *too fast*; Diana has *short* hair; if inflation continues *high* then it is *very unlikely* that there will be a *significant reduction* of taxes in the *near future*; the news are *unlikely* to be *true*; and so on, the words in italics denoting fuzzy attributes and concepts.

(c)   The third level is called fuzzy-perceptual level of processing. At this level, words, propositions and sentences of natural language are characterized by linguistic variables and clusters of propositions [Zadeh, 1973 and 1999] and they are mathematically represented by fuzzy sets and membership functions [Zadeh, 1965] along with fuzzy granules and generalized fuzzy constraints [Zadeh, 1996a and 1999]. The machine computations at this level use the principles of fuzzy logic and fuzzy constraint propagation to manipulate such complex symbols. In such a way, the *cognitive machine* mimics some of the levels of processing of the human mind.

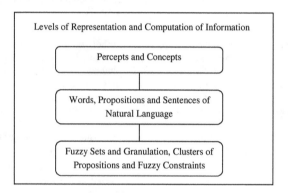

Figure 3.8. The Three Levels of Symbolic-Processing of the *Cognitive Machine*

3.5.4. Short-Term Memory

This memory simply stores fuzzy-perceptual symbols for a period of time given by KT_0. It works as an input and output device for the decision-making process block.

3.5.5. Categorization

Categorization provides humans with the ability to organize knowledge via the approaches to concept identification and hierarchy of classes. One of the benefits of categorizing objects is the reduction of complexity of the environment[30] [Reed, 1988].

To categorize is to group objects into classes in order to form concepts. Furthermore, these classes can be hierarchically organized into sub-ordinate and super-ordinate relations - i.e. some categories can contain other categories such as the category furniture contains chairs

---

[30] By classifying objects as being equivalent, humans respond to them in terms of their degree of membership into a class rather than as unique items.

and tables.

3.5.6. Concept Identification

Concepts are categories of physical, abstract and mental objects with common properties - like *red* and *green* are categories of colours, *short* and *tall* are categories of height, *small* and *big* are categories of size, *slow* and *fast* are categories of speed, *cheap* and *expensive* are categories of price, and so on. In fact, most of the concepts manipulated by the human mind have fuzzy boundaries [Bernstein *et al*, 1997; Lefrançois, 1995; and Zadeh, 2001] - i.e. the classification of an object into a category is a matter of degree and in this sense some objects have greater degrees of membership to a category than other objects.

In such a view, the theories of computing with words and computation of perceptions play an important part in concept identification for *cognitive machines*. Within such theories, concepts are described by words, propositions and sentences of natural language [Zadeh, 1996a, 1999 and 2001].

In the Figure 3.7, four blocks use concepts. They are: the long-term memory which comprises the representation of knowledge through concepts; the input transducer or fuzzifier which forms concepts (fuzzy-perceptual symbols) from input signals (sensations); the decision-making process which manipulates concepts by propagating them from premises to conclusions; and finally, the output transducer or defuzzifier which maps such concepts to signals compatible to the environment.

Similarly to the perspectives of Gestalt psychologists who proposed principles which describes how perceptual systems group sensations to form patterns [Bernstein *et al*, 1997], this research assumes that such principles are applicable to concept identification when classifying objects into categories. Some principles called "indistinguishable, similarity, proximity and functionality" have been proposed by Zadeh [2001]. He asserts that objects can be drawn together by using one of these principles to form fuzzy granules. In his theory, fuzzy granules are denotations of words, and words are labels of percepts.

Concepts can be defined as artificial and natural types [Bernstein, *et al* 1997].

- *Artificial Concepts*

Artificial concepts are characterized by classes of sharp boundaries. In such a type, an object is classified as member or non-member of a given class like in the ordinary set theory [Halmos, 1960]. Moreover, the discrimination of objects as members and non-members of a class is clearly defined according to a set of logical rules that comprise conjunctive, disjunctive, conditional or bi-conditional properties [Reed, 1988]. Such rules are similar to the propositions and principles of ordinary logic of predicate calculus [Luger and Stubblefield, 1998] and crisp-granular computation [Zadeh, 2001].

An example of artificial concept is the class of *small* numbers which are in between 0 and 10 inclusive: $\{0 \geq X \leq 10 \mid X \in R \text{ (Real)}\}$. In such a case, the conjunctive and conditional rule given by:

$$\text{if } X \text{ is } \geq 0 \text{ and } X \text{ is } \leq 10 \text{ then } X \text{ is } small \qquad (3.1)$$

is the rule that constrains X within the category of *small*. However, most of the concepts humans form and manipulate - like *small* - have no sharp boundaries and they are characterized by vagueness [Black, 1937 and 1963] and fuzziness [Zadeh, 1965]. Hence, the principles of ordinary sets theory and crisp computation do not provide the appropriate and necessary armament to represent and to manipulate such concepts.

Despite being used in laboratories of psychology for experimental research involving tasks such as the simulation of learning processes [Bernstein, *et al* 1997], artificial concepts of sharp boundaries fail to represent the complexity of real (natural) concepts.

- *Natural Concepts*

Natural concepts pervade our world and require complex representations which go beyond the limitations of ordinary sets theory and crisp granulation. Natural concepts have fuzzy boundaries because some of their members seem to be better examples of the category than others [Bernstein *et al*, 1997]. For instance, although a large range of stimulus input may be interpreted as being *green* (i.e. belonging to the category of *greenness*) some of that input will be interpreted as being more *green* and some as less *green*. Such categories are characterized by classes whose boundaries are not clearly defined and hence the transition from membership to non-membership of their members is gradual rather than abrupt. Examples include the classes of *short* and *tall*, *young* and *old*, *black* and *white*, *poor* and *rich*, *true* and *false*, *high* and *low* performances, etc.

In such a way, fuzzy sets theory[31] (along with fuzzy logic, computing with words and computation of perceptions [Zadeh, 2001]) can provide *cognitive machines* with the appropriate and necessary armament for the representation of natural concepts and categories through complex symbols by using the principles of:

- Fuzzy sets and membership functions [Zadeh, 1965; and Klir and Folger, 1992].
- Linguistic variables and fuzzy granulation [Zadeh, 1973 and 1996a].
- Generalized fuzzy constraints [Zadeh, 1999].

Figure 3.9 illustrates a mathematical representation of the fuzzy concept *young* via the principle of membership functions of fuzzy sets. In such an example, *young* is a linguistic value which represents the concept that constrains the variable age into such a category of *young*. Moreover, by considering that A denotes the fuzzy set *young*, $u \in U$ denotes the objects (u) in the universe of discourse (U) of age, and $\mu_A(u)$ represents the respective degree of membership of u in $A \subset U$, then the fuzzy set *young* can be symbolically equated as:

$$A = \{u, \mu_A(u) \mid \mu_A(u) \in [0,1], u \in U, A \subset U\} \qquad (3.2)$$

As much as u approaches 100 in the universe U, it assumes smaller degrees of membership within the class of *young* people. In such an example, the fuzzy set *young* is equivalent to a fuzzy-perceptual symbol.

---

[31] Fuzzy sets theory can also be regarded as an extension of ordinary sets theory since it provides additional mathematical principles for the representation of information in the form of more complex symbols. In the approaches to computing with words and perceptions [Zadeh, 1999] these symbols denote words and percepts.

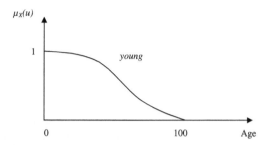

Figure 3.9. A Fuzzy-Perceptual Symbol representing the Concept of *Young*

It is important to realize that the imprecision and thus the fuzziness that is characteristic of natural concepts do not necessarily imply loss of accuracy or meaningfulness. It is for instance more meaningful and accurate to say that it is usually *warm* in the summer than to say that it is usually 25° C [Klir and Folger, 1992].

3.5.7. Long-Term Memory

This device stores the knowledge base of the *cognitive machine* and mental representations. It comprises:

- Concepts which are denotations of words, and conversely, words which are descriptions of concepts. Words are synonymous with linguistic variables which are mathematically represented by fuzzy sets [Zadeh, 1973, 1975 and 1976].

- Propositions P described in natural language which can be defined by fuzzy generalized constraints and represented via canonical forms. A proposition P can also be viewed as a constraint of a linguistic variable X to a particular category (e.g. P: X *isr* A, where *isr* is a variable copula which defines the way that the category A constrains the linguistic variable X) [Zadeh, 1999].

- Conditional statements (R) which describes the relations between concepts (linguistic values) in the antecedent and consequent (e.g. R: IF $X_{(i=1,...,n)}$ THEN $Y_{(j=1,...,M)}$ | $i,j \in N$ *integer*). The relation between the antecedent concepts can be defined by a t-norm $\cap$ and the relation between the consequent concepts can be represented by a s-norm $\perp$ [Dubois and Prade, 1985; and Nobre, 1997]. Moreover, the relation between the antecedent X and the consequent Y of such statements can be mathematically defined by fuzzy implications [Lee, 1990]. Conditional statements of this type can also be called fuzzy conditional rules [Zadeh, 1973].

- Clusters of propositions and conditional statements ($R_{(r=1,...,M)}$ | $r \in N$ integer) which can be defined as a set of fuzzy rules aggregated by an s-norm (e.g. $R_1 \perp R_2 \perp ... \perp R_M$) [Dubois and Prade, 1985; and Nobre, 1997].

The design of the knowledge base of the *cognitive machine* can be done by [Nobre, 1997

67

and Wang, 1994]:

- Using the experience, knowledge and heuristic rules of thumb of experts.
- Using computational programs for optimization, learning and automatic generation of conditional statements.

3.5.8. Decision-Making Process

This process manipulates concepts through the rules of inference in fuzzy logic [Zadeh, 1999]. In other words, it propagates fuzzy constraints from premises (antecedent concepts) to conclusions (consequent concepts).

The decision-making process associates fuzzy-perceptual symbols with the knowledge base stored in memory in order to make choices through approximate reasoning mechanisms.

The concept of approximate reasoning [Gupta and Sanchez, 1982; and Sanchez and Zadeh, 1987] is used here as synonymous with economic decision-making and satisfactory outcomes as defined in bounded rationality [Simon, 1997a] rather than the high costs and unrealistic view of optimal standards and pure rationality as employed in classical economics.

One of the most popular rules of inference in fuzzy logic is the compositional rule [Zadeh, 1996a]. To illustrate such a rule of inference, let us firstly define a set of fuzzy conditional statements and their aggregation given by:

$$R^{(r)} : \text{IF } x_1 \text{ isr } F_1^r \text{ AND IF } x_2 \text{ isr } F_2^r \text{ AND} \dots \text{AND IF } x_n \text{ isr } F_n^r \text{ THEN } y \text{ isr } G^r \qquad (3.3)$$

$$\mathbf{R} = \bigcup_{r=1}^{M} R^{(r)} \qquad (3.4)$$

where in equation (3.3), $R^{(r)}$ denotes a fuzzy conditional statement and $r=1,\dots,M$ is the total number of statements, and F_1^r and G^r denotes concepts (linguistic values) of the respective fuzzy variables $x_{(i=1,\dots,n)}$ and y. The logical connective AND is interpreted as a t-norm $\cap$ and THEN is implemented as an implication function.

In equation (3.4), $\mathbf{R}$ denotes the aggregation of the conditional statements given by the union operator $\mathbf{U}$ which also symbolizes a s-norm $\perp$.

Equations (3.3) and (3.4) summarize mental representations of knowledge in the long-term memory. The process of decision-making comprises the manipulation of such mental representations in order to compute an output B from an input fuzzy-perceptual symbol A. Figure 3.10 illustrates such a process and equation (3.5) resumes the calculus of decisions through the compositional rule of inference.

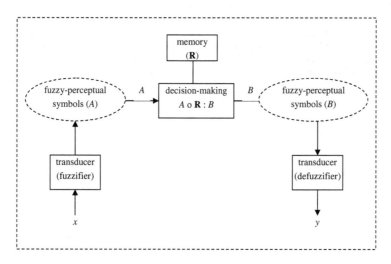

Figure 3.10. Decision-Making Process ($A \circ \mathbf{R} : B$)

$$A \circ \mathbf{R} = \mu_B(y) = \sup_{x \in U_{(i=1,\dots,n)}} [\mu_A(x) \cap \mu_\mathbf{R}(x, y)] \qquad (3.5)$$

where sup abbreviates supremum, $x = (x_1,\dots,x_n) \in U_{(i=1,\dots,n)}$ denote sensations (signals) which are mapped by the input transducer (fuzzifier) to a fuzzy-perceptual symbol A whose structure is characterized according to a fuzzy set. A denotes a word which is a description of a percept and thus a concept. B can be understood as a fuzzy-perceptual symbol calculated from equation (3.5) and $y \in V$ denotes the output signal after transduction (defuzzification).

3.5.9. Transduction: From Perceptual-Fuzzy Symbols to Signals

The (output) transducer is equivalent to a defuzzifier [Jager, 1995; Lee, 1990; Nobre, 1997; and Wang, 1994]. It maps fuzzy-perceptual symbols inferred from the decision-making process to signals compatible with the environment.

3.6. Summary

This chapter introduced the design of *cognitive machines*. It contributed with a methodology to bring selected technologies of machines (and artificial intelligence) closer to the discipline of cognition.

The cognitive processes associated with the *cognitive machine* were perception, attention and concept identification; short-term and long-term memory storage; representation and organization of knowledge via categorization and concept identification; and decision-making. Additionally, the disciplines selected to model and to govern the functioning of the *cognitive machine* were fuzzy sets theory, fuzzy logic, computing with words and computation of perceptions [Zadeh, 1999 and 2001].

The design comprised a framework of *cognitive machines* with the ability to manipulate percepts [Nobre and Steiner, 2003a]. Percepts and thus concepts (along with mental models) are represented by words, propositions and sentences of natural language. The ability of these machines to manipulate a percept provides them with higher levels of information-processing than other symbolic-processing machines; and according to the theory of levels of processing in cognition [Reed, 1988], these machines mimic (even through simple models) cognitive processes of humans [Nobre and Steiner, 2003a].

CHAPTER 4. ANALYSIS OF *COGNITIVE MACHINES* IN ORGANIZATIONS

4.1. Introduction

This chapter introduces the analysis of *cognitive machines* and perspectives about their participation in organizations. Therefore, it connects *cognitive machines* with the discipline of organizations.

The analysis of the *cognitive machines* comprises concepts of bounded rationality, economic decision-making and conflict resolution [Nobre and Steiner, 2003b]. From such an analysis this research advocates that these machines can be used to reduce or to solve intra-individual and group dysfunctional conflicts which arise from decision-making processes in organizations. Therefore, they can provide organizations with higher degrees of cognition, and consequently reduce the relative level of complexity of the environment. This chapter concludes by presenting perspectives about the work relationships between *cognitive machines*, their designer and the organization.

4.2. Capability Boundaries of *Cognitive Machines*

Bounded rationality and economic decision-making are characteristic processes of the human mind [Simon, 1997a]. Therefore, they are discussed in this section in order to understand some capability boundaries of *cognitive machines*.

4.2.1. Bounded Rationality and Economic Decision-Making

The fundamental premises about bounded rationality are [March, 1994]:

- Limitation of knowledge (or scarcity of information).
- And limitation of computational capacity (or limit of cognition).

In order to cope with such limitations, humans search for approximate and satisfactory solutions rather than optimal outcomes in their daily life [March and Simon, 1993] - where the term satisfactory is synonym for satisficing [Simon, 1997a]. The process of decision used by humans which lead them to satisfactory outcomes is called economic decision-making [Nobre and Steiner, 2003b] and it is synonymous with approximate reasoning [Sanchez and Zadeh, 1987]. Therefore, economic decision-making is concerned with economy in the processes of decision which result in satisfactory solutions rather than with processes of choice which search for optimal outcomes such as in neo-classical economics [Simon, 1997a]. Economic decision-making processes play an important part in the environments where information is scarce and fuzzy, and the cost of searching and computation of information is high.

Theories of choice that do not assume the preceding premises seem to be unrealistic and they cannot provide models of human cognition [Simon, 1997a].

4.2.2. Extending the Boundaries of Human Cognition with Machines

With the advent of computers and communication networks, along with the disciplines of operational research and management science, new technologies sough to extend the limits of rationality established by the cognitive boundaries of individuals and organizations [March and Simon, 1993; and Simon, 1982b]. A powerful combination of massive knowledge storage with high capability of symbolic-processing (and more specifically numerical and analytical computation) gave computers and thus information technology special places in organizations [Simon, 1977].

However, despite providing organizations with economic and cognitive contributions, computers and information technology have found serious limitations of applicability in those areas where problems and decisions require approximate rather than precise formulations. Such areas involve managerial roles and thus the management of decisions at the higher levels of the organization hierarchy.

The advancements of artificial intelligence in the period between the middle and the end of the 20th Century gave genesis to additional computational tools with the capability to solve classes of problems which could not be formulated before (e.g. chess and theorem proof). Nevertheless, the progress of artificial intelligence has been slow and limited in those areas where the formulation of problems falls into the category of fuzzy-granulation rather than crisp-granulation [Zadeh, 2001].

Most of the concepts that humans manipulate have fuzzy boundaries and the representation of these concepts requires new approaches to encapsulate them into more complex symbolic structures. According to the principle of a theory of levels of information-processing, humans can achieve higher levels of cognition when they reason with higher levels of symbolic representation such as words and sentences of natural language. In organizations, for instance, the higher the management level, the more complex are the concepts and mental models that humans reason with.

In such a way, further advancements in cognition and artificial intelligence research may lead researchers to engineer *cognitive machines* which combine the cognitive strengths of humans and computers.

On the one hand, people have a remarkable ability to reason with fuzzy concepts and to solve problems[32] through approximate and satisfactory solutions. They have a large long-term memory, but a very limited short-term memory. They also have limitations to reason with numerical and analytical representations of symbols.

On the other hand, computers are still poor at solving problems which require the formulation and manipulation of natural concepts along with approximate reasoning. They have a large memory and no distinction is needed between short and long-term memories. Hence, they overcome the limits of human short-term memory. Computer also overcomes the inability of humans at solving arithmetic and analytical problems [Simon, 1982b].

Figure 4.11 brings together the strengths of human cognition and computers to illustrate the abilities of a *cognitive machine*.

---

[32] Such problems involve tasks such as driving in city traffic, playing football and most of the management decisions at upper-levels in the organization hierarchy.

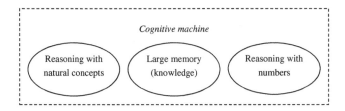

Figure 4.11. Abilities of the *Cognitive Machine*

4.2.3. Conditions in the Design of the *Cognitive Machine*

The design of the *cognitive machine* framework presented in Chapter 3 assumes similar conditions to bounded rationality and human decision-making processes [March and Simon, 1993 and Simon, 1997a]:

- Alternatives of choice are not simply given but they must be generated through a process of search.

- The probability distributions of outcomes are unknown and may be only estimated through high computational costs.

- Humans manipulate natural concepts [Bernstein, *et al* 1997]. Therefore, most of the uncertainty that pervades the alternatives and their consequences are classified into fuzziness rather than probabilistic uncertainty.

- Satisfactory strategies of choice and outcomes are preferable to maximization (the search for optimal solutions). The latter requires a higher cost of computation and may not represent a robust procedure [Zadeh, 1994 and 2001].

Therefore, in order to satisfy such conditions, the design of the *cognitive machines* assumes that:

- Alternatives (conditional statements or rules) which form the knowledge base of the machine are searched and generated by human experts or with the support of computational tools of adaptive and learning capabilities [Nobre, 1997 and Wang, 1994]. The process of search and generation of conditional statements is better described by a combination of the logic of appropriateness and the logic of consequences[33] [March, 1994; and Simon, 1982b]. Therefore, it involves experience, intuition, expertise along with calculation.

- The higher the number of alternatives, the higher is the number of rules; the higher number of rules, the higher is the completeness of the knowledge base [Nobre, 1997]; the higher the completeness of the knowledge base, the higher is the design complexity.

---

[33] In the logic consequences, actions are selected by evaluating their expected consequences for the preferences of the actor. It is related to the conception of calculation and analysis. In the logic of appropriateness, actions are matched to situations by means of rules. It involves conceptions of experience, roles, intuition and expertise [March, 1994; and March and Simon, 1993].

- Fuzziness is the type of uncertainty that pervades the alternatives (antecedents) and their consequences (conclusions). The knowledge base of the *cognitive machine* comprises antecedents and conclusions which describe relations between natural concepts. Such concepts have fuzzy boundaries. Therefore, fuzzy sets theory is a necessary tool for the representation of such concepts.

- A satisfactory and more robust strategy for decision-making is implemented through the principles of fuzzy logic, computing with words and computation of perceptions [Sanchez and Zadeh, 1987; and Zadeh, 1996a and 1999].

4.3. The Role of *Cognitive Machines* in Conflict Resolution

This section regards *cognitive machines* as decision-makers in organizations. It describes how such machines can contribute to improve processes of choice in the organization by reducing decision conflicts.

4.3.1. Decision Processes in Organizations

Organizations comprise several kinds of decisions [March and Simon, 1993]. However, this research is mainly concerned with decisions that influence and control the organization business and management. Such decisions are made by the participants within the organization and they involve conflicts (frictions).

4.3.2. Conflicts in Organizations

• *Constructive and Dysfunctional Conflicts*

Conflicts shape and affect the behaviour of individuals, groups and organizations [Daft and Noe, 2001]. They can be classified into constructive and dysfunctional conflicts.

On the one hand, constructive conflicts are classes of conflicts which contribute to improve the behaviour and performance of individuals, groups and organizations. On the other hand, dysfunctional conflicts are synonymous with obstacles which limit the action and performance of individuals, groups and organizations.

This research is concerned with dysfunctional conflicts which arise from decision-making processes in organizations.

• *Intra-Individual Conflict*

Processes of decision-making involve trade-offs among alternatives which are characterized by uncertainty, incomparability and unacceptability and hence they can lead organization participants to intra-individual conflict [March and Simon, 1993]. Such a kind of conflict arises in an individual mind and it also can emerge from the influence of others.

The Problem of Uncertainty: When considering models of rational choice and calculation which follow a logic of consequences, uncertainty means that the probability distributions of outcomes are unknown [March, 1994; and Simon, 1982b]. This research assumes either: that such probabilities cannot be estimated or they can be calculated only with unrealistic costs of computation.

The Problem of Incomparability: It means that the individual (participant in the organization) cannot recognize a most preferred alternative. It can happen for instance when the individual has to decide between two alternatives with the same label such as *good*.

The Problem of Unacceptability: It means that the most preferred alternative as identified by the individual does not satisfy standard criteria.

- **Group Conflict**

In addition to the factors that lead participants to intra-individual conflict, members of groups in organizations can differ in their perceptions, values and culture, needs and goals [Daft and Noe, 2001]. Hence, they can disagree in their decisions causing group conflict [March and Simon, 1993]. This kind of conflict arises from differences between the choices made by distinct participants within the organization.

- **Relations between Conflicts and Bounded Rationality**

The intra-individual and group conflicts which arise in organizations are mainly influenced by lack of information and uncertainties, and most importantly by cognitive limitations. In such a way, these conflicts cannot be solved by incentive and reward systems. Such cognitive and information constraints are synonymous with bounded rationality [March, 1994; March and Simon, 1993; Simon 1997a and 1997b]. However, this research asserts that *cognitive machines* can be used to reduce or to solve such conflicts.

4.3.3. *Cognitive Machines* in Conflict Resolution of Decisions

This subsection proposes principles to support the assertion that *cognitive machines* can reduce or solve intra-individual and group dysfunctional conflicts in organizations.

- **Resolution of Intra-Individual Conflict**

The reduction of the frictions in intra-individual conflicts can be achieved by providing the organization with means to cope with uncertainty, incomparability and unacceptability factors.

A Solution to Uncertainty: According to the theory of natural concepts proposed in the literature of cognition [Bernstein et al, 1997], most of the concepts which humans manipulate have fuzzy boundaries. Hence, fuzziness is the principal kind of uncertainty that the *cognitive machine* must deal with and manage during task execution and decision-making. Fuzzy concepts can be represented through complex symbols whose structure is properly defined via fuzzy sets along with the principles of fuzzy logic, computing with words and computation of perceptions [Zadeh, 1999 and 2001].

A Solution to Incomparability: Instead of identifying a most preferred conditional rule (or an alternative) the *cognitive machine* fires (or selects) a set of rules according to a criterion - for instance, the rules whose value is greater than 0 - and then it unifies (or aggregates) the fired set of rules through the application of one of the operators in fuzzy logic. Such operators comprise s-norm and union [Jager, 1995]. Therefore, aggregation of preferences and rules is applied rather than the selection of only one alternative.

A Solution to Unacceptability: This is avoided by using criteria of design during the specification of the *cognitive machine* knowledge base. For such a purpose, the criteria of completeness must be applied in order to guarantee that for each state there is an associated output [Jager, 1995 and Nobre, 1997].

75

- *Resolution of Group Conflict*

The absence of intra-individual conflict reduces group conflict, but it does not extinguish the problem since it is not a sufficient condition. By assuming such an absence, this subsection discusses the additional agents of group conflict and it proposes solutions to solve it.

Group conflict also arises from divergences of opinions of the participants in a group which can be a consequence of the differences in their needs, goals, values and perceptions. On the one hand, such a type of conflict could be reduced by equalizing the participants' perceptions, opinions and knowledge. On the other hand, it could be solved through a methodology which supports the integration of the participants' perceptions, opinion and knowledge. The latter is the selected approach used in this research to justify the way in which *cognitive machines* reduce (or solve) group conflicts. It consists of integrating the participants' perceptions, opinion and knowledge in a common knowledge base (device storage or memory).

The design of the *cognitive machines* introduced in Chapter 3 comprises the specification of a knowledge base through commonsensical expertise – i.e. different experts express their perceptions, opinions and knowledge in the form of words and sentences of natural language which take the form of linguistic rules or conditional statements [Zadeh, 1973 and 1996a]. Such rules may also be generated automatically, modified and improved through the principles of adaptive and learning systems [Wang, 1994]. During the functioning of the *cognitive machine*, the activation of rules and their aggregation represent an integrated and common sense process which takes into account the perceptions, opinion and knowledge of different experts.

Therefore, decision-making processes are automated through the rules of inference of fuzzy logic [Zadeh, 1996a, 1999 and 2001]. Moreover, intra-individual and group mental models are represented through a set of fuzzy propositions and fuzzy conditional statements (rules) - which can be mathematically defined through fuzzy generalized constraints [Zadeh, 1999].

- *Networks of Cognitive Machines in Conflict*

When operating in networks, similarly to multi-agent systems [Weiss, 1999], *cognitive machines* can get in conflict during negotiation. This topic will be investigated in further work.

4.4. *Cognitive Machines*, Designers and Organizations: Relationships

This section describes relationships between the *cognitive machine*, its designer and the organization within which the machine participates. It proposes definitions of designers of *cognitive machines* and their duties; consciousness of the *cognitive machine* in the organization; and work responsibilities of the *cognitive machine*, its designer and the organization.

4.4.1. On Designers of *Cognitive Machines*

Definition 4.4.1.1: The designer (which involves the manufacturer) of a *cognitive machine* is the person (or entity - organization) responsible for the cognitive abilities and the behaviour of the machine. Such a kind of designers can have independent legal identity which enables them to make contracts and to seek court enforcement of those contracts if necessary.

Definition 4.4.1.2: Designers can use technologies for automatic design and generation of *cognitive machines*. Genetic programming for instance is a paradigm which has provided a profound impact on the design of software programs capable of generating tangible replicas and with enough ability to perform at least similar functions [Koza, J.R. 1992]. Such technologies are classified as artificial designers and they cannot ask for, nor answer, a formal contract. Therefore, their first designer (a person or an organization) must do.

4.4.2. On Consciousness of *Cognitive Machines* in Organizations

This research does not intend to investigate whether *cognitive machines* are able to be conscious[34] or whether they can be self-aware of their roles and tasks in organizations. It would require a profound analysis of the subject of consciousness in cognitive science and artificial intelligence research [Haikonen, 2003]. Nevertheless, this research puts forward a definition of what machine consciousness means to organizations.

Definition 4.4.2.1: Machine consciousness represents the awareness of its designer in relation to the cognitive processes and abilities that the machine carries on during task execution.

4.4.3. On Responsibility: The Designer, the Machine and the Organization

What is the relationship between a *cognitive machine*, its tasks and its roles in the organization? And what is the work relationship between the *cognitive machine*, the machine designer and the organization?

Definition 4.4.3.1: The work relationship between the machine designer and the organization can be regularized by a contract which makes explicit the cognitive abilities of the machine; the tasks that the machine can perform within the organization; the roles that the machine fulfils in the organization; and also the designer and the organization attestation (or signatures).

Definition 4.4.3.2: The organization is responsible for the assignment of roles to the *cognitive machine*, and the machine is responsible for the roles it serves the organization. However, the machine designer and the organization are the main parts responsible for the machine results and performance. If the machine exhibits deviant behaviour during task execution or performance below specified criteria, then the contract between the organization and the machine designer is the object of analysis and judgement.

---

[34] Consciousness concerns mental states of being aware of oneself and one's environment. It assumes the awareness of one's own mental processes, thoughts, feelings and perceptions. Consciousness states can vary from deep sleep to alert wakefulness [Bernstein, D.A. et al 1997].

4.5. Summary

This chapter complemented the preceding one by presenting the analysis of *cognitive machines* and perspectives about their participation in organizations. It contributed by connecting *cognitive machines* with the discipline of organizations.

The analysis of the *cognitive machine* comprised concepts of bounded rationality, economic decision-making and conflict resolution [Nobre and Steiner, 2003b]. Such an analysis indicates that these machines can be used to reduce or to solve intra-individual and group dysfunctional conflicts which arise from decision-making processes in organizations. Therefore, they can provide organizations with higher degrees of cognition.

This chapter concluded by presenting work relationships between the *cognitive machine*, its designer and the organization. As a participant within the organization, the *cognitive machine* must fulfil roles as designated by the organization. Additionally, a contract can form the relationship between the designer of the *cognitive machine* and the organization in order to ensure responsibility.

FINDINGS OF PART II

Part II introduced concepts about organization cognition. It is based on the premise that *cognitive machines* can improve the degree of cognition of the organization, and also upon the proposition that an increase in the degree of organization cognition reduces the relative levels of uncertainty and complexity of the environment with which the organization relates. Part II comprised Chapters 2, 3 and 4.

Chapter 2 introduced a methodology in order to support the choice of organization design strategies which increase the degree of cognition of the organization and reduce the relative level of complexity and uncertainty of the environment. From this methodology, the technology and the participants in the organization were chosen as the elements of design because they comprise *cognitive machines*. It also introduced definitions about organizations, the environment and relations between them as viewed throughout this research. Such definitions include concepts for intelligence, cognition, autonomy and complexity of organizations and machines along with environmental complexity. From such definitions it was established that:

- Theorem 2.5.1.1: The technology of *cognitive machines* increases the level of complexity of the organization (and thus the degree of cognition of the organization), and it relatively reduces the level of environmental complexity (and uncertainty) that the organization confronts.

Chapter 3 introduced the design of the organization elements chosen in the organization design methodology of Chapter 2. Particular attention was given to the design of *cognitive machines*. They were chosen in order to increase the degree of cognition of the organization and thus to increase the organization ability to process information and to make decisions.

The design of *cognitive machines* contributed with a methodology to bring selected technologies of machines (and artificial intelligence) closer to the discipline of cognition. The cognitive processes associated with the *cognitive machine* were perception, attention and concept identification; short-term and long-term memory storage; representation and organization of knowledge via categorization and concept identification; and decision-making. The disciplines selected to model and to govern the functioning of the *cognitive machine* were fuzzy sets theory, fuzzy logic, computing with words and computation of perceptions [Zadeh, 1999 and 2001].

The design methodology provided a framework of *cognitive machines* with the ability to manipulate percepts [Nobre and Steiner, 2003a]. Percepts and thus concepts (along with mental models) are represented by words, propositions and sentences of natural language [Zadeh, 2001]. The ability of these machines to manipulate a percept provides them with higher levels of information-processing than other symbolic-processing machines; and according to the theory of levels of processing in cognition [Reed, 1988], these machines mimic (even through simple models) cognitive processes of humans [Nobre and Steiner, 2003a].

Chapter 4 introduced the analysis of *cognitive machine* and perspectives about their participation in organizations. The analysis comprised concepts of bounded rationality, economic decision-making and conflict resolution. Such an analysis indicates that these machines can be used to reduce or to solve intra-individual and group dysfunctional conflicts which arise from decision-making processes in organizations. Therefore, they can provide organizations with higher degrees of cognition. Chapter 4 concluded by presenting work relationships between the *cognitive machine*, its designer and the organization. As a participant of the organization, the *cognitive machine* must fulfil roles as designated by the organization. Additionally, a contract can form the relationship between the designer of the *cognitive machine* and the organization in order to ensure responsibility.

PART III: STUDY OF ORGANIZATIONS

"...as the complexity of a system increases, our ability to make precise and yet significant statements about its behaviour diminish until a threshold is reached beyond which precision and significance (or relevance) become almost exclusive characteristics. It is in this sense that precise quantitative analyses of the behaviour of humanistic systems are not likely to have much relevance to the real-world societal, political, economic, and other types of problems which involve humans either as individual or in groups."

Principle of Incompatibility stated by Lotfi A. Zadeh [1973]

This research supports the statement of Zadeh [1996b] with the perspective of humanistic systems defined by social systems – i.e. systems whose behaviour is preponderantly influenced by emotions, cognition and social networks. Therefore, besides quantitative analysis, Part III uses qualitative and computational approaches to the study of organizations which involve the representation of percepts, concepts and mental models through words and propositions (linguistic conditional statements) of natural language [Zadeh, 1999].

Part III provides real data from the market and it comprises Chapter 5 only.

Chapter 5 presents an industrial case within NEC [Nobre and Nakasone, 1999; Nobre and Volpe, 1999 and 2000; Nobre *et al*, 2000; and Nobre and Steiner, 2001a]. This case is concerned with practices of organizational learning and it involves two complementary activities: - process and technology change management.

The first activity is concerned with the implementation of a continuous process improvement model in one of the industrial plants of NEC - the organizational model is called The Capability Maturity Model (CMM) proposed by the Software Engineering Institute of the Carnegie Mellon University. This activity involves measures of organization process maturity and organization cognition levels along with performance indexes. These measures and indexes are used as quantitative indicators of organizational learning and process improvement.

The second activity is concerned with the pilot application of a *cognitive machine* in the adaptive learning cycle of the Radio Engineering Department of NEC do Brasil S.A. It involves tasks of analysis, decision and management control of the performance indexes of five successive, discrete large-scale software projects which comprises tangible (budget, schedule and requirement completeness) and intangible outputs (customer satisfaction and project process quality). The design of the *cognitive machine* is reinforced with a set of criteria along with qualitative (i.e. the phase plane approach) and quantitative (i.e. mathematical and convergence) analysis.

Results of this industrial case showed that the Capability Maturity Model (CMM) provided the organization with improvements in its process maturity level. However, despite improving project development and management processes at the organization technical level, the progress of the CMM in those areas at higher hierarchical layers of the organization (such as managerial and institutional levels) was slow and poor due to lack of commitment to the CMM at these higher layers.

The growth of organization performance was indicated through measures of customer satisfaction and project process quality computed by the *cognitive machine*. Such a growth of performance was associated with improvements in the CMM maturity level (which is synonymous with the level of organization process maturity) and organization cognition.

Moreover, these results could be measured in two ways. Firstly, on an integer scale [1,5] which indicates the degree of organization cognition correlated with the level of organization process maturity. Secondly, on a real scale [0,10] which indicates the level of organization performance correlated with the level of organization process maturity.

CHAPTER 5. AN INDUSTRIAL CASE

5.1. Introduction

This chapter presents studies about an organization[35] (NEC) through the approaches of participant observation, computational modelling along with qualitative and quantitative analysis. Such approaches were selected according to the criteria introduced in the Appendix A and their definitions are briefly reviewed in the following.

- *Participant Observation*

Participant observation concerns the practical experience of the author in NEC do Brasil S.A during the period between 1997 and 2000. During that period, NEC do Brasil S.A. was investing in new process improvement models and the author was a full-time engineer within the Radio Engineering Department where he was fulfilling the roles of a Master in Electronics, Software and Quality Engineering.

- *Computational Modelling*

Computational modelling supports the design and implementation of a *cognitive machine* in a computer using software and programming tools. Such a *cognitive machine* plays an important part in tasks of analysis, decision and management control of the performance of five successive large-scale software projects. This machine was designed and implemented in the Radio Engineering Department of NEC do Brasil S.A. in 1998 by the author. This adventure yielded him the 1998 Industrial Director Award in the category of Software Process Improvement and Quality in the organization of study (NEC).

- *Qualitative and Quantitative Research*

Qualitative and quantitative analysis (or mathematical research) is used in this case to support the design of the *cognitive machine* and the improvements achieved in the levels of the organization's cognition, process maturity and performance.

5.2. The Organization of Study: NEC / NDB

The Radio Engineering Department of NEC do Brasil S.A. is the core organization (or unit) of study. NEC do Brasil S.A. (http://www.nec.com.br) is a subsidiary of NEC (http://www.nec.com) which is a Japanese enterprise spread worldwide in the business market of communications, multimedia, information technology, semiconductors, computers along with other technological fields. NEC do Brasil S.A. (hereafter abbreviated by NDB) is a telecommunications company which acts in the Brazilian along with the Latin American market. Its main industrial plant is located in the city of Sao Paulo, in Brazil. Table 5.1 presents some of the data about NDB such as size, wealth and age [Nobre and Volpe, 1999 and 2000].

---

[35] This chapter regards the organization as defined in Part II. Additionally, it considers that the organization can be regarded as a unit (such as a division or department) within a company or other entity which manages a set of projects. These projects within the organization share common management and policies [Paulk *et al*, 1994].

NDB provides customers with telecommunications products and services whose performance and quality criteria are highly dependent on software systems such as real time and embedded applications along with Telecommunications Management Networks (TMN) of International Telecommunication Union (ITU) - ITU is the United Nations Specialized Agency in the field of telecommunications [ITU-T, 2000].

Table 5.1. NEC do Brasil S.A. (NDB)

Foundation	26th/11/1968
Liquid Wealth	US$ 1,154.00 (milhions) / in 1998
Market products	Telecommunications systems (TMN – Telecommunications Management Networks, Wireless, Radio, Transmission and Switching systems)
Number of employees	2,932 (in December 1998)
Address	Rod. Presidente Dutra, km 214, Guarulhos-Sao Paulo, 07210-902, Brazil
Competitors	Ericsson, Siemens, Motorola, Nortel-Lucent

5.3. The Context and Methodology of Study

5.3.1. Context: Process and Technology Change Management

During 1996 NEC do Brasil S.A. (NDB) started to invest in a specific and international program for software improvement which involved activities of process and technology change management. This program complemented and supported its organizational focus on total quality management (TQM) and customer satisfaction. After researching the market in partnership with The University of Sao Paulo (http://www.usp.br), NDB chose the Capability Maturity Model (CMM) as its main direction model to develop guidelines for improving its organization's software process [Nobre and Volpe, 1999 and 2000].

5.3.2. The Methodology of Study

As explained in Appendix A and shortly described in the introduction of this Chapter, participant observation was the main methodology of study of this industrial case and it was supported with computational modelling and mathematical research.

- *Data Source and Collection*

Between 1997 and 2000 the author was employed within the Radio Engineering Department of NDB as a full-time engineer. Hence, this Chapter describes his practical experience within NDB and it uses data collected from internal reports of the organization and published in worldwide congresses by him and other researchers within NDB [Nobre and Nakasone, 1999; Nobre and Volpe, 1999; Nobre and Volpe, 2000; and Nobre et al, 2000].

The internal reports and publications include:

(i) Data from the development of five successive large-scale software projects such as budget (C), schedule (T) and number of requirements (R). The recording of these data was required according to the procedures of software project planning, tracking and oversight of the Radio Engineering Department of NDB. Such data are further used in this investigation as input factors for the computation of performance indexes of the software projects.

(ii) Feedback from project and product clients relating to customer satisfaction (CS) - a performance index of the organization in this investigation. Such feedback was collected by the Quality and Radio Engineering Departments of the Radio Systems Division during internal meetings with representatives of the clients of the software products along with quality auditors and engineers of NDB.

(iii) Feedback from members (engineers and managers) within the Radio Engineering and Quality Departments of the Radio Systems Division of NDB about project process quality (PPQ) - a second performance index of the organization in this investigation.

- *The Roles of the Participant (Author) in NDB*

The successive roles and responsibilities of the author within NDB between 1997 and 2000 were:

(i) Development of software programs for Telecommunications Management Networks in the Radio Engineering Department (i.e. software and electronics engineering tasks during 1997).

(ii) Research and development of software engineering, software processes and software quality assurance practices for the Radio Engineering Department of NDB (i.e. software and quality engineering tasks during 1997 and 1998).

(iii) Research and development of software process improvement policies for the whole organization NDB and coordination of software quality assurance practices for the Radio Engineering Department (i.e. management of software and quality engineering tasks during 1999 and 2000).

5.4. The Capability Maturity Model (CMM): Overview

The Capability Maturity Model (hereafter abbreviated by CMM) is a software process maturity framework proposed by the Software Engineering Institute of the Carnegie Mellon University [Paulk *et al*, 1994]. The CMM comprises the application of process management concepts of Total Quality Management (TQM) to software. It provides recommendations and guidelines for software engineering and management practices along with an evolutionary path of five levels of maturity to improve the software process capability of organizations. It also provides methods to appraise and to assess the maturity level of the organization's software process.

Recent research reported by the Software Engineering Institute of the Carnegie Mellon University has shown that the number of organizations using the CMM has reached 1,500 organizations worldwide [SEI-CMU, 2004]. A summary of important benefits of the CMM to organizations can be found in [Herbsleb *et al*, 1994; and Paulk and Chrissis, 2000]. Among the assessed improvements and measures include return on investment (ROI), gain per year in productivity, reduction of schedule to develop software systems, product quality and defect reduction.

5.4.1. Process Capability, Maturity and Performance of the Organization

Process capability, maturity and performance are distinct but complementary definitions within the CMM [Paulk *et al*, 1994]. These terms are defined in the following since they are often used throughout this chapter.

- ***Organization Process Capability***

This is concerned with the expectation of results and thus predictability - i.e. with the amount of expected results that can be achieved by following a process.

- ***Organization Process Maturity***

This is concerned with the level of specification of a process and it comprises the extent to which a particular process is explicitly designed, defined, institutionalized, managed, controlled and effective. Process maturity provides organizations with the potential for capability growth.

- ***Organization Process Performance***

This is concerned with actual results achieved by following a process.

5.4.2. Mature *vs.* Immature Organizations

Before describing the levels of maturity of the CMM, this subsection differentiates immature from mature software organizations.

- ***Immature Organizations***

Immature organizations are characterized by low levels of predictability[36], control[37] and effectiveness[38] [Paulk *et al*, 1994]. Their process is poorly understood and their ability to solve problems depends on the particular skills of an expert during crisis. Such organizations routinely exceed schedules and budgets because they are not based on realistic planning. Consequently, it is highly probable that immature organizations cannot satisfy their goals, targets and strict criteria of customer satisfaction, quality, time, cost, etc. Nevertheless, immature organizations can achieve successful results when they pay a high price for heroic experts.

- ***Mature Organizations***

In contrast, mature organizations are characterized by high levels of predictability, control and effectiveness. Their process is clearly defined and their projects are based on realistic planning. In such a type of organizations, activities are carried out according to a planned process and hence it is most probable that they can satisfy their goals, targets and criteria.

---

[36] Improvements in predictability reduce the difference between target and actual results across projects.

[37] Improvements in control reduce the variability of actual outcomes around target results.

[38] Improvements in effectiveness improve target results and thus provide the organization with the ability to satisfy stricter criteria - such as lower cost, shorter schedule (time), and higher productivity and quality.

Additionally, they comprise continuous process improvement practices which provide them with the abilities to learn by creating and managing knowledge [Argote, 1999].

5.5. Maturity Levels of the CMM

The CMM framework involves five levels of maturity as illustrated in Figure 5.1. They form an evolutionary path to support continuous process improvement and organizational learning practices [Paulk *et al*, 1994].

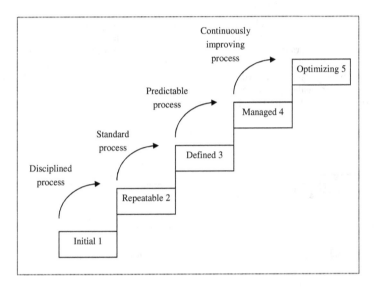

Figure 5.1. The CMM Maturity Levels

5.5.1. Level 1: The Initial Level

At level 1, the organization process is characterized as ad hoc and unstable. Success in organizations at level 1 depends on the competence and heroics of their participants and cannot be repeated unless the same competent individuals are assigned to the next projects. Capability is a merit of the individuals, but not of the organization.

At level 1, the organization process is poorly defined and amorphous – like a black box – and visibility into the process is limited. Hence, managers have difficult to identify the status of the activities of the projects and customers can only assess whether the product satisfy requirements when it is delivered. Figure 5.2 illustrates an organization process at level 1.

Figure 5.2. Visibility into the Organization Process at Level 1

5.5.2. Level 2: The Repeatable Level

At level 2, the organization process is specified along with basic management practices which comprise requirements management, project planning, tracking and oversight, subcontract management, quality assurance and configuration management. Success is achieved with discipline and experience accumulated with similar projects. The organization process is specified and divided into successive black boxes as illustrated in Figure 5.3. Hence, managers have better visibility into the process. They can track the status of projects and products at transition points (milestones) of the organization process.

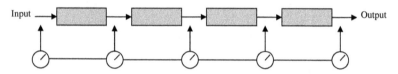

Figure 5.3. Visibility into the Organization Process at Level 2: Repeatable

5.5.3. Level 3: The Defined Level

At level 3, the organization process is defined, standardized, institutionalized and used across the organization. It comprises engineering and management practices. This process is referred by the CMM as the organization's standard software process (OSSP). The projects of the divisions and units of the organization tailor the OSSP and develop their own software process. A tailored process is called a project's defined software process (PDSP). Figure 5.4 illustrates the relation between the OSSP and the PDSP of three divisions of an organization.

Moreover, at level 3, managers have clear visibility into the organization process and into the internal structure of the boxes (stages) which form the project's process. Hence, the relation between management and engineering activities are understood, and the status of the project and products can be accurately updated and controlled. The integration of different project's defined software processes (PDSP) across the organization provides the organization's standard software process (OSSP) with new practices and thus with organizational learning. The best practices within the PDSP are recognized, standardized and institutionalized within the OSSP. Figure 5.5 illustrates such a process at level 3.

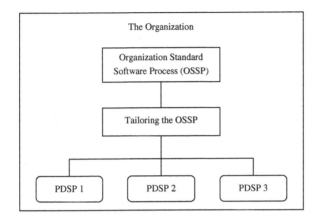

Figure 5.4. Relation of the OSSP and PDSP at Level 3

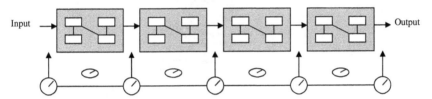

Figure 5.5. Visibility into the Organization Process at Level 3: Defined

5.5.4. Level 4: The Managed Level

At level 4, the organization establishes quantitative criteria and measures for software products and processes. Management and engineering decisions are based on quantitative measurements and they provide the organization with improvements in predictability, control and effectiveness. Hence, managers have visibility into the organization and project's processes and also access to measurements such as quality and productivity. Figure 5.6 illustrates the organization process at level 4.

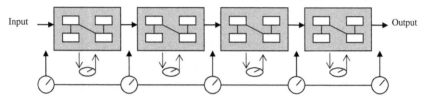

Figure 5.6. Visibility into the Organization Process at Level 4: Managed

5.5.5. Level 5: The Optimizing Level

At level 5, the organization focuses on continuous process improvement principles for defect prevention along with the management of process and technology change. Managers are able to predict and to control the implications of change and innovation of processes, products and technology for the organization and projects quantitatively. Technology change and process improvement are planned and managed as ordinary activities in the organization. Figure 5.7 illustrates the organization process at level 5.

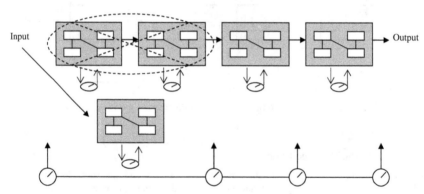

Figure 5.7. Visibility into the Organization Process at Level 5: Optimizing

5.6. Degree of Organization Cognition *vs*. Level of Process Maturity

As defined in Chapter 2, organization cognition comprises a set of processes which together form the organization cognitive system. Such a set of processes is synonymous with the human cognitive system. It was also defined in Chapter 2 that the degree of cognition of the organization is associated with the level of elaboration and integration of its cognitive processes.

This section assumes that the degree of cognition of the organization can be associated with and thus defined as contingent upon the level of maturity of the organization process. Therefore, this chapter assumes by definition that:

Proposition 5.6.1: The higher the level of organization's process maturity, the higher is the degree of organization cognition.

Proposition 5.6.1 plays an important part in this chapter since it associates degree of organization cognition with measures of process maturity. In such a way, the degree of organization cognition can be associated with one of the five maturity levels of the CMM. Therefore, the degree of organization cognition can be measured in the integer interval [1,5], where 1,…,5 represent the levels of process maturity of the CMM-based organization. Figure 5.8 illustrates the direct relation between levels of process maturity and degrees of organization cognition. The dotted arrow indicates a direct relation between these variables.

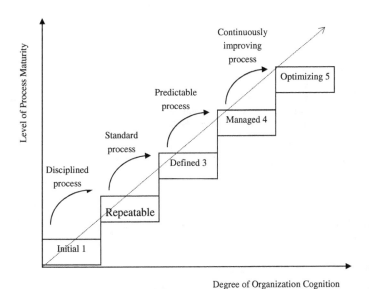

Figure 5.8. Level of Process Maturity *vs.* Degree of Organization Cognition

5.7. The CMM in NEC / NDB

A simplified structure of the industrial plant of NEC do Brasil S.A. (NDB) is illustrated in Figure 5.9. It comprises four divisions: - Radio, Transmission, Switching and Wireless Communications Systems.

Such Divisions shared common organization processes whose procedures and practices were based on ISO 9000, ISO 14000 and Lean manufacturing concepts [Nobre and Volpe, 1999]. Additionally, they had their own software process and in 1996 they started to invest in the implementation of the Capability Maturity Model (CMM).

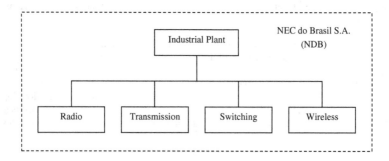

Figure 5.9. Divisions of the Industrial Plant of NEC do Brasil S.A. (NDB)

In December of 1997 and during the year of 1998, such divisions were assessed by an authorized lead evaluator of the Software Engineering Institute of the Carnegie Mellon University. They were officially recognized as satisfying the criteria of the CMM level 2. Therefore, from the year of 1999 NDB started to invest towards the achievement of the CMM level 3. This achievement was unsuccessful until the end of 2000 due to budget constraints and changes in the Brazilian market of telecommunications caused by national privatizations of governmental telecommunications companies [Nobre and Volpe, 2000].

5.8. A *Cognitive Machine* as a Participant in NEC / NDB

5.8.1. Aim

The pilot application and thus the participation of a *cognitive machine* in the Radio Engineering Department of NDB comprised analysis, decision and management control tasks about the performance of large-scale software projects of the Radio Engineering Department of NDB.

Moreover, this application was aimed to:

(a) Provide managers and stakeholders with information about the organization performance, including information about process and product quality along with customer satisfaction.

(b) Support managers in the analysis, decision and control of the organization performance.

(c) Reduce intra-individual and group dysfunctional conflicts which arise from managerial decisions.

5.8.2. Historical Facts

As a participant of NDB during the period between 1997 and 2000, the author designed and implemented a pilot *cognitive machine* in the Radio Engineering Department.

The idea of the design and implementation of a *cognitive machine* in NDB was paved by three main factors:

(a) In 1998 NDB was in the apogee of its software process improvement program.

(b) NDB was providing incentives through the Industrial Director Award for organization process improvement and quality innovators.

(c) As a participant of NDB during the period between 1997 and 2000, the author held a master of sciences (MSc) background in those technologies and disciplines of *cognitive machines* as introduced in Chapter 3.

5.8.3. The Radio Engineering Department of NDB: The Core Unit of Study

The Radio Engineering Department of NDB is the core unit of the organization of study. Therefore, it is briefly described in this subsection.

During the period between 1997 and 2000, the Radio Systems Division of NDB comprised several departments among those of product planning, engineering, manufacturing, process and product quality assurance, and product test and field implementation.

Within the Engineering Department of the Radio Systems Division of NDB was included the Telecommunications Management Networks (TMN) Section which was the principal unit responsible for the development of large-scale software systems of International Telecommunication Union standard (ITU) - the United Nations Specialized Agency in the field of telecommunications [ITU-T, 2000]. Figure 5.10 illustrates the structure of the Radio Systems Division.

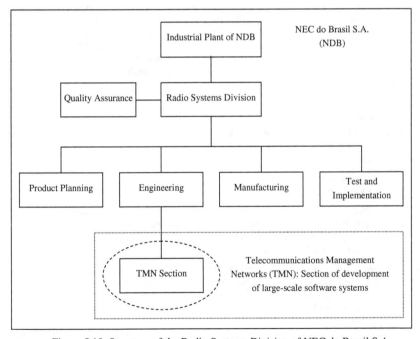

Figure 5.10. Structure of the Radio Systems Division of NEC do Brasil S.A.

5.8.4. Scope: Management Control of the Organization Performance

NDB, like other business organizations, comprises the concept of open-rational systems. Therefore, it needs an effective and efficient management control system to drive its state variables towards its goals.

Management control is a continuous process used to improve the capability of the organization to carry out its activities towards the achievement of its goals [Anthony *et al*, 1984]. Such a process comprises: the design of a control system structure; the definition of the state variables of the organization; the gathering of quantitative and qualitative information about these variables and their storage in the organization database (memory); the analysis of this information in order to derive conclusions about the organization performance; decision-making and control actions to improve the organization performance.

This research is concerned with such a perspective of management control.

5.8.5. Motivations: *Cognitive Machines* in Organization Analysis, Decision and Control

The development of fuzzy sets theory [Zadeh, 1965] and fuzzy logic [1973] was widely motivated by the need for an alternative and complementary approach, of mathematical and computational background, to the analysis of complex systems – such as those systems found within the category of social sciences – i.e. systems whose behaviour is determined by human emotion, cognition and relationships [Karwowski and Mital, 1986]. Most recently, theories of computing with words [Zadeh, 1996a] and computation of perceptions [Zadeh, 1999 and 2001] were proposed to provide the literature with new approaches to the representation and manipulation of mental models. From such a perspective, this section takes advantage of the elements of these disciplines to design a *cognitive machine* whose behaviour is based on the skills and knowledge of managers and engineers within the Radio Engineering Department of NDB [Nobre *et al*, 2000]. Such an expertise comprises technical and managerial knowledge about software projects, software process and product quality, organization process improvement and management control systems.

5.8.6. Challenge

The provision of feedback information to stakeholders about the performance of an organization encompasses three key problems [Nobre and Nakasone, 1999; and Nobre and Steiner, 2001a]:

(i)   The first problem is concerned with the concept of performance and the identification of the state variables which influence it in the organization.

(ii) The second problem concerns the vagueness [Black, 1937 and 1963] and fuzziness [Klir and Folger, 1992; and Zadeh, 1965] inherent in the concept of performance. Hence, the representation of the qualitative aspects of this concept through a precise quantitative symbol becomes a difficult task.

(iii) The third problem concerns the intra-individual and group dysfunctional conflicts which arise from decisions involving the concept of organization performance. Such decisions comprise subjective judgments and alternatives which are characterized by uncertainty, incomparability and unacceptability. These alternatives include state variables in the premises and in the conclusions of propositions; such state variables influence the organization performance and they comprise natural concepts which are inherently characterized with fuzzy boundaries.

5.8.7. Solution

To contribute to the solution of the previous problems, this research introduces an approach whose features are presented in the following.

(i) Organization performance is defined as contingent upon indexes of customer satisfaction (CS) and project process quality (PPQ). Such performance indexes are defined as contingent upon project cost (C), project schedule (T) and product requirements completeness (R). C, T and R are named organization or project performance factors; they represent state variables and their relations form the premises to the conclusion about CS and PPQ. The concepts of C, T and R are defined in the Appendix E.

(ii) The concepts of the organization performance's indexes (CS and PPQ) and factors (C, T and R) are mathematically represented through membership functions of fuzzy sets theory [Zadeh, 1965]. Therefore, the fuzzy boundaries inherent in these concepts take the form of complex symbols whose structures are represented by fuzzy sets.

(iii) Decision-making processes about the organization performance are automated through the rules of inference of fuzzy logic [Zadeh, 1996a, 1999 and 2001]. Moreover, intra-individual and group's mental models are represented through a set of fuzzy propositions and fuzzy conditional statements - which can be mathematically defined by fuzzy generalized constraints [Zadeh, 1999]. Fuzzy conditional statements also play an important part in this application by providing measures (f) to map the factors $X=[C, T, R]$ to the indexes $Y=[CS, PPQ]$:

$$f: X \rightarrow Y \qquad (5.1)$$

Before introducing the design of the *cognitive machine* and its analysis, the next subsection presents a framework of a management control system used in the analysis, decision and control tasks of the organization performance. Additionally, it describes the role of the *cognitive machine* into such a framework.

5.8.8. A Framework of a Management Control System

Such a framework is illustrated in Figure 5.11. It comprises a managerial cycle with principles of feedback control systems and continuous process improvement [Nobre *et al*, 2000; and Nobre and Steiner, 2001a]. Additionally, it can be regarded as an adaptive learning cycle with operation principles of single-loop and double-loop learning [Daft and Noe, 2001].

The organization is regarded as the Radio Engineering Department of NDB and its process represents the guidelines, procedures and policies for software project development and management of the TMN Section of the Radio Engineering Department. Therefore, the term organization performance is used as synonymous with the performance of the management of the software projects of the TMN Section.

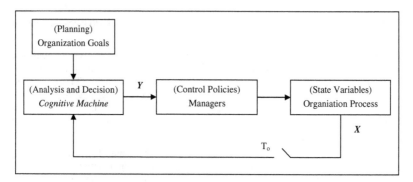

Figure 5.11. Management Control System of the Organization Performance

Put shortly, this framework involves the activities of:

(i) Planning: This is concerned with the design of the organization goals and it includes the specification of the organization performance criteria, indexes and factors along with measures. A performance criterion represents the boundaries of a target to be achieved. Indexes (Y) are representations of the organization performance and in this investigation they comprise customer satisfaction (CS) and project process quality (PPQ): Y=[CS, PPQ]; indexes are contingent upon performance factors. Factors (X) are represented by the state variables of the organization process and they comprise project cost (C), project schedule (T) and product requirements completeness (R): $X = [C, T, R]$; during the stage of planning their values are estimated through statistical techniques and experience. The factors C, T and R are defined according to Appendix E. Measures are functions (f) used to map factors (state variables X) to performance indexes (Y) according to equation 5.1.

(ii) Sampling T_o: This is concerned with the collection of qualitative and quantitative information about the state variables (X) of the organization process. X comprises C, T and R and they are synonymous with the performance factors of the organization (Appendix E). The sampling time (T_o) represents the period of time for the collection of new information about X. In this investigation T_o is defined as equal to the planning schedule. Therefore, information about the actual cost (C_A) and actual requirements completeness (R_A) of a project is collected from the organization process when T_o is equal to the planning schedule.

(iii) Analysis and Decision: This is concerned with the evaluation and computation of the organization performance. It maps the factors C, T and R to the performance indexes CS and PPQ. The task of analysis and decision is performed by the *cognitive machine*. It computes the CS and PPQ indexes from the actual cost (C_A) and requirements completeness (R_A) information which is sampled at time T_o from the organization process. The planning stage provides the *cognitive machine* with information about C_o and R_o (which represent the planning cost and the planning requirements completeness as defined in the Appendix E). Figure 5.12 illustrates the input-output information with the *cognitive machine*. The actual schedule (T_A) is assumed as equal to the sampling time T_o and thus it is not considered as a state variable of computation.

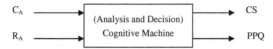

Figure 5.12. Input-Output Information with the *Cognitive Machine*

(iv) Control Policies: This is concerned with the directions and actions that managers take over the organization process in order to improve performance indexes. It comprises process and technology change management. Such control policies are mainly based on the guidelines and recommendations of the Capability Maturity Model (CMM).

5.8.9. Review of the Data Collection Process

• *Successive Software Projects*

The industrial case presented in this chapter is mainly concerned with five successive, discrete software projects of the Telecommunications Management Networks (TMN) Section of the Radio Engineering Department of NEC do Brasil S.A. (NDB). These projects were developed between 1997 and 2000 and their levels of complexity were similar. Therefore, according to the guidelines of the CMM maturity level 2, the organization can achieve success and then it can repeat it through discipline and experience accumulated with such similar projects.

• *Participants in the Software Projects*

The participants (staff) of the TMN Section were one manager and ten engineers of software and electronics engineering background – including the author. They were trained in the CMM guidelines and framework, and they remained together between 1997 and 2000 when they were working on the development of TMN software projects whose process maturity was officially acknowledged by evaluators of the Software Engineering Institute of Carnegie Mellon University as CMM Maturity Level 2 [Nobre and Volpe, 1999].

• *Choice of the Performance Factors and Indexes*

The choice of the factors (state variables of the software process) and indexes of performance of the software projects was a group decision by the TMN Section staff with additional feedback from the clients of the software projects and experts of the Quality Department of the Radio Systems Division.

• *Source of the Data about the Software Projects*

The data relating to the cost (C), requirements completeness (R) and schedule (T) of the five TMN software projects were collected from internal reports, database and publications of the Radio Engineering Department of NEC do Brasil S.A. [Nobre and Nakasone, 1999; and Nobre *et al*, 2000].

5.9. Level of Organization Performance vs. Process Maturity

This section assumes that organization performance can be associated with and thus defined as contingent upon the level of maturity of the organization process. Therefore, it assumes by definition that:

Proposition 5.9.1: The higher the level of organization's process maturity (and thus the higher the degree of organization cognition), the greater is the chance of the organization to exhibit high levels of performance.

Proposition 5.9.1 plays an important part in this chapter since it associates level of organization performance with measures of process maturity and thus organization cognition. In such a way, the performance indexes defined through customer satisfaction (CS) and project process quality (PPQ) can be used as additional elements to measure the progress of the CMM in the organization of study.

The next section introduces the design of the *cognitive machine*.

5.10. Design of the *Cognitive Machine*

5.10.1. Structure

The structure of the *cognitive machine* along with its functional blocks is presented in Figure 5.13. This structure is tailored from the general framework presented in Figure 3.7 and its decision-making process works according to the Figure 3.10. Additionally, this structure is based on the classic configuration of fuzzy logic controllers [Lee, 1990; Nobre, 1997; and Wang, 1994].

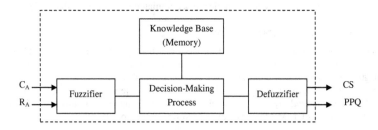

Figure 5.13. The Structure of the *Cognitive Machine*

5.10.2. Criteria of Design

The *cognitive machine* was designed to satisfy the set of criteria presented in the following [Nobre, 1997; and Nobre *et al*, 2000]. Such criteria involve terminologies and concepts which are defined throughput the literature of fuzzy systems [Klir and Folger, 1992; Pedrycz and Gomide, 1998; and Wang 1994].

Criteria C1: The fuzzy sets of the input variables satisfy the definition of fuzzy numbers and fuzzy partitions.

Criteria C2: The fuzzy sets of the output variables satisfy the definition of fuzzy numbers and their centre contains only one element.

Criteria C3: The rule base (or the set of fuzzy conditional statements) satisfies the definition of strict completeness.

Criteria C4: The AND logical operator is implemented as the algebraic product.

Criteria C5: The OR logical operator is implemented as the bounded sum.

Criteria C6: The implication function satisfies a fuzzy conjunction and it is implemented as the algebraic product.

Criteria C7: The singleton fuzzifier is the operator defined to the fuzzification of input variables.

Criteria C8: The centre average defuzzifier is the operator defined to the defuzzification of output variables.

The definition of criteria C1 to C8 was motivated because they provide the designer with a *cognitive machine* of simply structure and algorithm which can be easily implemented in computers and investigated through mathematical approaches to convergence and stability analysis [Nobre, 1997; and Wang, 1994]. Moreover, criterion C3 guarantees that for all states assumed by the input variables X = [C, R], there exists an output state Y = [CS, PPQ] which can be computed by the decision-making process [Jager, 1995; and Nobre, 1997].

The specification of the functional blocks and parameters of the *cognitive machine* are presented in the following.

5.10.3. Description of Percepts via Words and Linguistic Variables

Linguistic variables involve descriptions of percepts via words and fuzzy granules which are synonymous with linguistic values of a variable. The input C and R and the output variables CS and PPQ have their linguistic values presented in the following. Their granularity and concepts were defined according to the experience of the *cognitive machine* designer (i.e. the author) along with the knowledge of the manager and other engineers of the TMN Section of the Radio Engineering Department of NDB.

- *Input Variables*

$$C = [cheap, \ not \ so \ cheap, \ expensive]$$
$$R = [empty, \ almost \ empty, \ partial, \ almost \ full, \ full]$$

- *Output Variables*

$$CS = [very \ low, \ low, \ medium, \ high, \ very \ high]$$
$$PPQ = [really \ bad, \ very \ bad, \ bad, \ moderate, \ good, \ very \ good, \ really \ good]$$

5.10.4. Representation of Concepts via Membership Functions of Fuzzy Sets

Membership functions of fuzzy sets are used to represent linguistic variables and words (and thus percepts and concepts) through complex symbols of mathematical background[39]. The representations of the input C and R and output variables CS and PPQ via fuzzy sets are depicted in Figures 5.14 and 5.15. They were defined according to criteria C1 and C2. Their triangular shape and universe of discourse were specified according to the experience of the *cognitive machine* designer and also using the knowledge of the manager and other engineers of the TMN Section of the Radio Engineering Department of NDB.

In Figure 5.14, μ_{CA} and μ_{RA} denote the degrees of membership of C_A and R_A in their respective fuzzy sets, where: μ_{CA} and $\mu_{RA} \in [0,1]$; the words *cheap, not so cheap* and *expensive* are labels of the fuzzy sets which characterize the concept of cost (C) defined on the universe of discourse C_A; and the words *empty, almost empty, partial, almost full* and *full* are labels of the fuzzy sets which characterize the concept of requirements completeness (R) defined in the universe of discourse R_A.

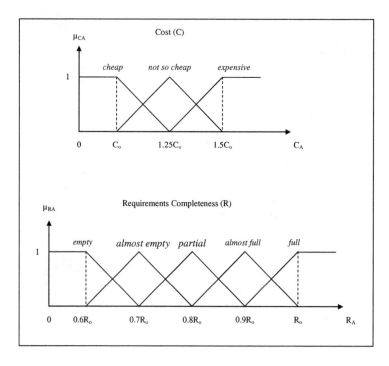

Figure 5.14. Fuzzy Sets of the Organization's Performance Factors: State Variables

---

[39] Linguistic variables are synonymous with fuzzy variables when they are represented by fuzzy sets.

In Figure 5.15, μ_{CS} and μ_{PPQ} denote the degrees of membership of CS and PPQ in their respective fuzzy sets, where: μ_{CS} and $\mu_{PPQ} \in [0,1]$; the words *very low*, *low*, *medium*, *high* and *very high* are labels of the fuzzy sets which characterize the concept of customer satisfaction (CS) defined in the universe of discourse CS; and the words *really bad*, *very bad*, *bad*, *moderate* and *good*, *very good* and *really good* are labels of the fuzzy sets which characterize the concept of project process quality (PPQ) defined in the universe of discourse PPQ.

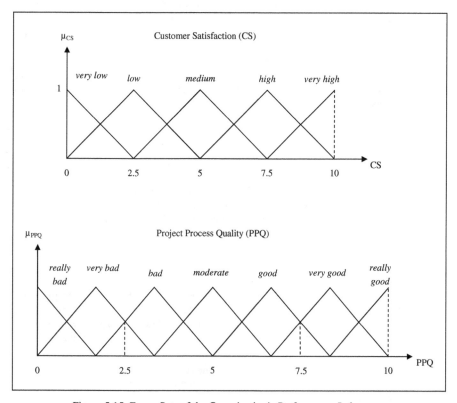

Figure 5.15. Fuzzy Sets of the Organization's Performance Indexes

5.10.5. Representation of Mental Models via Fuzzy Conditional Statements

Fuzzy conditional statements describe relations between two or more fuzzy variables in the form of:

IF A THEN B: $(A \rightarrow B)$ (5.2.)

where A and B denote fuzzy variables; THEN can be defined as a fuzzy implication (A→B); and the symbol → denotes an implication function [Zadeh, 1973].

Fuzzy conditional statements can be classified as a particular approach to perception-based system modelling [Zadeh, 2001]. In this chapter, this approach is used to design and to represent propositions and mental models which describe the relations between input (factors) and output (indexes) variables of the management control system of the organization performance. Such mental models comprise the experience and the perception of the participants (engineers and the manager) of the TMN Section of the Radio Engineering Department of NDB, to the organization behaviour and to the concept of organization performance.

- **_Design of the Fuzzy Rule Bases_**

The set of fuzzy conditional statements comprises two fuzzy rule bases which satisfy the criterion C3. The first one is concerned with the analysis and conclusion about the customer satisfaction (CS) which is inferred from product requirements completeness (R). Therefore, it describes relations between R and CS:

$$\text{Fuzzy Rule Base 1: IF R THEN CS: (R} \rightarrow \text{CS)} \qquad (5.3)$$

The second rule base is concerned with the analysis and conclusion about the project process quality (PPQ) which is inferred from project cost (C) and product requirements completeness (R). Therefore, it describes relations between C, R and PPQ:

$$\text{Fuzzy Rule Base 2: IF C AND IF R THEN PPQ: (C AND R} \rightarrow \text{PPQ)} \qquad (5.4)$$

where AND satisfies the criterion of design C4 and the implication function → satisfies C6.

The fuzzy rule base 1, given by equation (5.3), does not include project cost (C) because this variable is not a matter for the customer. On the other hand, C is a control variable to projects' managers and organization's stakeholders. Hence, C is included in the fuzzy rule base 2, given by equation (5.4).

The set of fuzzy conditional statements of the fuzzy rule bases 1 and 2 are represented in Figures 5.16 and 5.17 respectively. The aggregation of the statements of each fuzzy rule base satisfies the criterion of design C5.

Customer Satisfaction (CS) Mental Models					
R	_empty_	_almost empty_	_partial_	_almost full_	_full_
CS	_very low_	_low_	_medium_	_high_	_very high_

Figure 5.16. Fuzzy Rule Base about Customer Satisfaction: R → CS

In Figure 5.16, the cells above the axis (arrow) contain the linguistic values of R and those cells below it contain the linguistic values of CS. Each pair of cells constituted by the linguistic values of R and CS which are opposite to each other and separated by the axis, forms one fuzzy conditional statement. Such a rule base comprises a set of five fuzzy conditional statements and they are described according to Appendix F. As an example, the first pair of cells whose linguistic values are in the front of R and CS forms one fuzzy conditional statement defined as:

$$\text{IF R is } empty \text{ THEN CS is } very\ low \qquad\qquad (5.5)$$

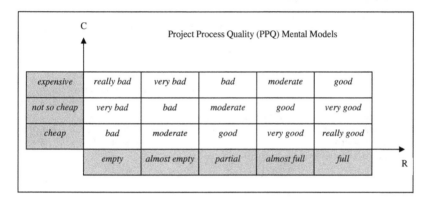

Figure 5.17. Fuzzy Rule Base about Projects' Process Quality: (C AND R) → PPQ

In Figure 5.17, the cells on the left side of the vertical axis contain the linguistic values of C and those cells below the horizontal axis contain the linguistic values of R. The other cells, located in between the C and R axes (in the first quadrant) contain the linguistic values of PPQ which represent conclusions of the fuzzy conditional statements. Such a rule base comprises a set of fifteen fuzzy conditional statements and they are described according to Appendix F. As an example, when the variables C and R assume the respective linguistic values of *not so cheap* and *full*, then PPQ assumes the linguistic value of *very good*. Such a fuzzy conditional statement is defined as:

$$\text{IF C is } not\ so\ cheap \text{ AND R is } full \text{ THEN PPQ is } very\ good \qquad (5.6)$$

5.10.6. Decision-Making Process via the Compositional Rule of Inference

The compositional rule of inference of fuzzy logic [Zadeh, 1973 and 1999] is the mechanism used to implement the decision-making process of the *cognitive machine*. This mechanism is described in Chapter 3. Put shortly, it manipulates concepts (and thus percepts) by propagating them from premises (antecedents of fuzzy conditional statements) to conclusions. It can also be defined as a mechanism to reason with linguistic representations of mental models.

5.11. Analysis of the *Cognitive Machine*

The approaches to qualitative and quantitative analysis presented in this section represent simplified versions of the methodology proposed in [Nobre, 1997].

The qualitative analysis is concerned with the design of the linguistic rules of the *cognitive machine* and thus it provides the designer with a methodology to improve the *cognitive machine*'s behaviour and performance. This approach is based on the concept of a phase plane whose state space is described and represented through linguistic values.

The quantitative analysis is concerned with the description of the linguistic rules of the *cognitive machine* through analytical equations. It provides the designer with a methodology to study some mathematical properties of the *cognitive machine* such as stability analysis [Nobre, 1997].

5.11.1. Qualitative Analysis

Figure 5.18 illustrates a phase plane of one dimension only which characterizes the relations between the linguistic variables (R and CS) and their respective linguistic values. The dotted arrows indicate the directions of increasing in the linguistic values of R and CS. It can be observed that the higher the linguistic value of R, the higher is the linguistic value of CS, because the higher the completion of the product's requirements at time T_0, the higher is the customer satisfaction [Nobre *et al*, 2000].

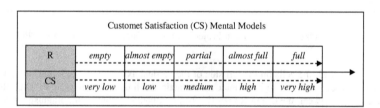

Figure 5.18. Analysis of Customer Satisfaction: R *vs.* CS

Figure 5.19 illustrates a phase plane of two dimensions which characterizes the relations between the linguistic variables (C, R and CS) and their respective linguistic values. The dotted arrows indicate the directions of increasing in the linguistic values of PPQ. It can be observed that the higher the linguistic value of R, the higher is the linguistic value of PPQ, because the higher the completion of the product's requirements at time T_0, the higher is the projects' process quality. Moreover, Figures 5.19 shows that the higher the linguistic value of C, the lower is the linguistic value of PPQ, because the higher the project cost at time T_0, the lower is the projects' process quality [Nobre *et al*, 2000].

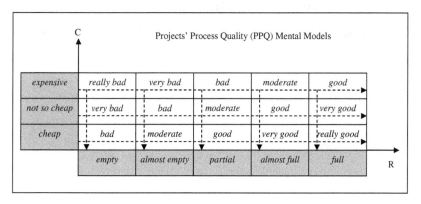

C	Projects' Process Quality (PPQ) Mental Models				
expensive	*really bad*	*very bad*	*bad*	*moderate*	*good*
not so cheap	*very bad*	*bad*	*moderate*	*good*	*very good*
cheap	*bad*	*moderate*	*good*	*very good*	*really good*
	empty	*almost empty*	*partial*	*almost full*	*full*

Figure 5.19. Analysis of Projects' Process Quality: (C AND R) *vs.* CS

5.11.2. Quantitative Analysis

This subsection investigates the boundaries and convergence of the output variables of the *cognitive machine* for the state space defined to its input variables [Nobre *et al*, 2000]. It starts by proposing the following:

Theorem 5.11.2.1: The application of criteria C1 to C8 results in a *cognitive machine* (or system) whose output variable is defined by:

$$y(t) = \sum_{r=1}^{M} \alpha_r . u_r \qquad (5.7)$$

where

$$\alpha_r = \prod_{i=1}^{N} \mu(x_i(t)) \qquad (5.8)$$

$y(t)$ denotes an output variable calculated at time t after the operation of defuzzification; M denotes the number of fuzzy rules; u_r denotes the centre of the fuzzy sets of the output variable; $i=1,...,N$ is the number of input variables $x_i(t)$; and $\mu(x_i(t))$ denotes membership functions of $x_i(t)$.

Proof 5.11.2.1: The proof is found in [Jager, 1995; and Nobre, 1997].

Therefore, the output variables of the *cognitive machine* - i.e. customer satisfaction (CS) and projects' process performance (PPQ) - can be derived from Theorem 1.

- *Analytical Calculus of Customer Satisfaction*

Theorem 5.11.2.2: According to Theorem 5.11.2.1, customer satisfaction (CS) can be defined as:

$$CS(t) = \sum_{r=1}^{5} \alpha_r . c_r \qquad (5.9)$$

where

$$\alpha_r = \mu_{R(t)} \qquad (5.10)$$

CS(t) denotes customer satisfaction calculated at time *t* after the operation of defuzzification; *r* =1,...,5 denotes the number of fuzzy rules; c_r denotes the center of the fuzzy sets of CS; *R(t)* denotes the value of product's requirements completeness (R) at time *t*; and $\mu_{R(t)}$ represents the membership function of *R(t)*.

Proof 5.11.2.2: Theorem 5.11.2.2 is a particular application of Theorem 5.11.2.1. Therefore, the proof of Theorem 5.11.2.2 is derived from Theorem 5.11.2.1.

- *Analytical Calculus of Projects' Process Quality*

Theorem 5.11.2.3: According to Theorem 5.11.2.1, project process quality (PPQ) can be defined as:

$$PPQ(t) = \sum_{r=1}^{15} \alpha_r . p_r \qquad (5.10)$$

where

$$\alpha_r = \mu_{C(t)} . \mu_{R(t)} \qquad (5.11)$$

PPQ(t) denotes projects' process quality calculated at time *t* after the operation of defuzzification; *r* =1,...,15 denotes the number of fuzzy rules; p_r denotes the center of the fuzzy sets of PPQ; *C(t)* and *R(t)* denotes the values of project cost (C) and product's requirements completeness (R) at time *t*; and $\mu_{C(t)}$ and $\mu_{R(t)}$ represent the membership functions of *C(t)* and *R(t)*.

Proof 5.11.2.3: Theorem 5.11.2.3 is a particular application of Theorem 5.11.2.1. Therefore, the proof of Theorem 5.11.2.3 is derived from Theorem 5.11.2.1.

- *Boundaries and Convergence Analysis*

Theorem 5.11.2.4: Given that the input variables (C and R) assume any real value in their respective universes of discourse, Theorems 5.11.2.2 and 5.11.2.3 guarantee that the outputs variables (CS and PPQ) of the *cognitive machine* converge to real values which are bounded to the interval [0,10] defined to their respective universes of discourse.

Proof 5.11.2.4: The proof is demonstrated in Appendix G.

5.12. Data and Results of the Industrial Case

5.12.1. Software Projects and Process of the Organization

This section is concerned with the results of the participation of the *cognitive machine* within the management control system of the Radio Engineering Department of NEC do Brasil S.A. (NDB) in the period between 1998 and 2000 [Nobre and Nakasone, 1999; and Nobre *et al*, 2000].

The role of the *cognitive machine* comprised the computation of the organization software projects' performance indexes - i.e. customer satisfaction (CS) and projects' process quality (PPQ) - from the state variables (factors) of the organization process - i.e. project cost (C), product requirements completeness (R) and project schedule (T=T_o) - where T_o denotes the sampling time which is equal to the planning schedule.

The organization process is concerned with a set of engineering and management practices defined according to the CMM maturity level 2 in order to support the development of successive software projects of the Telecommunications Management Networks (TMN) Section of the Radio Engineering Department of NEC do Brasil S.A. (NDB). Therefore, focus is directed upon data of software process, projects and products.

5.12.2. Data Gathering of TMN Software Projects

The data presented in this subsection was collected from internal reports, database and publications of the Radio Engineering Department of NEC do Brasil S.A. and they are concerned with a set of five successive large-scale software projects of telecommunication management networks as developed in the TMN Section of NDB during the years of 1997, 1998 and 1999. The software projects are enumerated chronologically to their development and they comprise a set of planning data (T_o, C_o and R_o) and a set of actual data (T_A, C_A and R_A) which were sampled at time $t = T_A = T_o$.

Table 5.1 contains the set of planning data of the five TMN software projects which are abbreviated by $SP_{(i=1,...,5)}$; Table 5.2 contains the set of actual data of these projects; and Table 5.3 contains the organization performance indexes (CS and PPQ) which were computed by the *cognitive machine* from the information about the planning and actual data of the software projects. The symbol OP in the Table 5.3 denotes Organization Performance and it was calculated by the arithmetic average between the performance indexes CS and PPQ.

It is important to understand that CS, PPQ and OP are performance indexes of TMN software projects which represent the organization of study. However, other indexes could be defined to represent the performance of the organization.

107

Table 5.2. Planned Data of the TMN Software Projects

Projects	Co (millions US$)	Ro	To (years)
SP1	0.2	50	0.4
SP2	0.5	100	0.9
SD3	1	150	1
SP4	1.2	180	1.5
SP5	2	300	2

Table 5.3. Actual Data of the TMN Software Projects

	C_A (millions US$)	R_A	T_A (years)
SP1	0.3	40	0.4
SP2	0.65	85	0.9
SP3	1	135	1
SP4	1.44	180	1.5
SP5	1.9	300	2

Table 5.4. Organization's Performance Indexes

	CS	PPQ	OP
SP1	5	2	3.5
SP2	6.25	2.8	4.52
SP3	7.5	5	6.25
SP4	10	6	8
SP5	10	10	10

5.12.3. Data Analysis

The graphical results of Table 5.3 are illustrated in Figure 5.20, where the scale [0,10] denotes measures of customer satisfaction(CS), projects' process performance (PPQ) and organization performance (OP).

The graphical results show that as we move from project SP_1 to SP_5, the indexes CS, PP and OP grow within the scale [0,10] until they reach the maximal planning criterion given by 10. From another point of view, these results indicate that the organization control policies are contributing to improve the organization performance indexes (PP, CS and OP). Such improvements can be attributed to the engineering and management practices provided with the CMM guidelines.

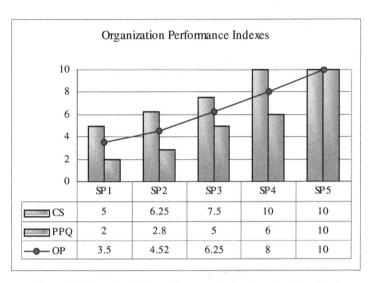

	SP 1	SP 2	SP 3	SP 4	SP 5
CS	5	6.25	7.5	10	10
PPQ	2	2.8	5	6	10
OP	3.5	4.52	6.25	8	10

Figure 5.20. Organization's Performance Indexes (CS, PPQ and OP)

The table in the following presents analysis of the results of Table 5.3 by comparing the planning and actual data of the software projects sampled at time $t = T_o$; where $SP_{(i=1,...,5)}$ enumerates the TMN software projects; and the symbols ↑ and ↓ denote up and down respectively.

Table 5.5. Analysis of the Organization's Performance

Software Projects	Analysis
SP1	C_A is 50% ↑ of C_o and R_A is 80% of R_o
SP2	C_A is 30% ↑ of C_o and R_A is 85% of R_o
SP3	C_A is equal to C_o and R_A is 90% of R_o
SP4	C_A is 20% ↑ of C_o and R_A is 100% of R_o
SP5	C_A is 5% ↓ of C_o and R_A is 100% of R_o

It can be observed that the actual cost (C_A) of the TMN software project SP_1 was 50% above the planning cost (C_o) and that the actual requirements (R_A) which were successfully implemented represents 80% of the total of the planning requirements (R_o). On the other hand, the actual cost (C_A) of the TMN software project SP_5 was 5% below the planning cost (C_o) and the actual requirements (R_A) which were successfully implemented represents 100% of the total of the planning requirements (R_o).

5.13. Findings at the Technical, Managerial and Institutional Levels

This section presents a postscript about NDB at different levels of analysis[40]. It represents the author's findings when he was playing the roles of:

- Software engineer in the Radio Engineering Department of NDB.
- Software quality assurance coordinator of the Division of Radio Systems and also of the whole NDB.
- Designer of policies of software process improvement of NDB.
- Researcher of the Software Engineering Process Group (SEPG) of NDB.

5.13.1. Levels of Analysis of the CMM in the NDB

- *The Technical Level*

The technical level of interest is concerned with the development of software projects which involve products, processes, technology and people of each division of NDB.

- *The Managerial Level*

The managerial level is concerned with the integration of the software processes of the Divisions of Radio, Transmission, Switching and Wireless Communications Systems of NDB. This integration is a requirement of CMM level 3.

- *The Institutional Level*

The institutional level is concerned with the alignment of the CMM goals at the technical and managerial levels with those at the organizational (and environmental) level. This level involve practices of organization design such as organizational strategy analysis which encompasses the review of products and services to be provided, the markets to be served and the value to be offered to the customers of NDB.

5.13.2. The Bottom-Up Strategy for Process Improvement

The idea of implementation of the CMM in the NDB was paved from the technical level by engineers and lower-level managers in the organization hierarchy. Then, this project was approved by higher-level managers at the managerial and institutional levels of the organization (NDB).

Nevertheless, despite being approved by managers at the managerial and institutional levels, the CMM project and its technical goals were not well aligned with those at the higher levels of the organization. In such a way, the acceptance of the idea and the approval of the CMM project were not encompassing practices of institutionalization at the higher levels of the organization. Areas such as marketing and organizational strategy were not aligned with the CMM goals.

As a result, the CMM was a successful project at the technical level since it guided the organization software projects to a continuous process improvement path for a period

---

[40] Such levels of analysis are defined in the Subsection 2.6.2.

110

characterized by a stationary environment. On the one hand, it was stationary because since its foundation, NDB had the ability to hold its main customers (Brazilian governmental telecommunications companies) for almost 30 years. Additionally, the software projects developed at the divisions and departments of NDB had similar requirements and thus similar levels of complexity to previous projects. Hence, they could be well developed with the guidelines of the CMM process maturity level 2 which provide repeatability in such circumstances. In such a way, success could be repeated with similar software projects. On the other hand, a stationary environment was characterized by the inability of NDB to attract and to expand its frontiers to new customers and markets.

5.13.3. The Effects of the Dynamics of the Environment of NDB

However, in August of 1995 and in July of 1997, the national congress of Brazil approved new constitutional laws which allowed the Brazilian government to begin a new era of privatization of the Brazilian market of telecommunications with concessions to worldwide competitors. This process was completed in July of 1998 after the privatization of EMBRATEL which used to be the biggest Brazilian enterprise in the area of telecommunications [ANATEL]. This privatization yields the Brazilian government more than 22 billions of Real (which is the Brazilian currency[41]).

After the year of 2000, NDB started to pass through a difficult economic situation with loss of some of its main customers caused by economic and political transformations of the environment. Hence, NDB was forced to re-design its social structure and goals which affected its participants (employees) drastically. Between 1998 and 2002, the number of participants (employees) of NDB was reduced by 90% approximately.

5.14. Summary

Chapter 5 presented an industrial case through the approaches to participant observation, computational modelling and qualitative and mathematical analysis. The organization of study was the Radio Engineering Department of NEC do Brasil S.A. which is a subsidiary of the Japanese enterprise NEC. This case involved organizational learning concepts and it comprised two complementary activities: - process and technology change management.

The first activity was concerned with the implementation of an organizational process improvement model in the industrial plant of NEC do Brasil S.A. (NDB), located in Sao Paulo, Brazil. This model is The Capability Maturity Model (CMM) of the Software Engineering Institute of the Carnegie Mellon University.

The CMM provided the organization with new engineering and management policies for continuous process improvement and organizational learning practices. It was defined that the degree of cognition of the organization can be associated with and thus contingent upon the level of process maturity of the organization. Therefore, improvements in the level of process maturity could be associated with improvements in the degree of cognition of the organization (and vice-versa).

Additionally, Chapter 5 showed that improvements in the performance of the organization were correlated with improvements in the level of organization process maturity.

---

[41] Today, one British Pound is equivalent to five Real approximately – i.e. a ratio of 1:5.

Therefore, improvements in the level of organization performance and organization cognition can be associated with improvements in the level of organization process maturity. Such improvements could be measured in two ways. Firstly, on an integer scale [1,5] which indicates the level of process maturity associated with the degree of organization cognition. Secondly, on a real scale [0,10] which indicates the level of organization performance (defined by indexes of customer satisfaction and projects' process quality) correlated with the level of organization process maturity.

The second activity was concerned with the application of a *cognitive machine* in the adaptive learning cycle (or management control system) of the Radio Engineering Department of NEC do Brasil S.A. The design of the *cognitive machine* was reinforced with a set of criteria along with qualitative and quantitative analysis. Such an analysis supported the design of the linguist rule bases of the *cognitive machine*. Additionally, it provided proofs about the convergence of the output variables of the *cognitive machine* – i.e. the convergence of the output variables to a bounded and real value contained withinthe interval [0,10] for all the values of the state space defined and bounded to the input variables.

The role and participation of the *cognitive machine* in the proposed management control system involved tasks of analysis, decision and management control of the performance of large-scale software projects. Such tasks included the computation of the organization performance indexes - defined as customer satisfaction (CS) and projects' process quality (PPQ) - from information of project cost (C) and product requirements completeness (R) sampled at time T_o from the organization process. The data gathered from the organization was concerned with five large-scale software projects developed in the Telecommunications Management Network (TMN) Section of the Radio Engineering Department of NDB during the years of 1997, 1998 and 1999.

The data analysis and results obtained from the management control system of the software projects indicated that investments in process and technology benefited the organization with improvements in the level of organization performance and in the associated degree of organization cognition. Therefore, the results supported the proposition that the higher the level of organization's process maturity, the higher is the degree of organization cognition and the level of organization performance. Such results satisfied Proposition 5.6.1.

Additionally, it was observed by the author that the participation of the *cognitive machine* in the management control system of the performance of the Radio Engineering Department of NDB contributed to:

- Provide managers and stakeholders with information about the organization performance; including information about process quality along with customer satisfaction.

- Support managers in the analysis and control of the organization performance.

- Reduce intra-individual and group dysfunctional conflicts which arise from decision-making processes in management control tasks.

Finally, the author summarized some of his findings about NEC do Brasi S.A. (NDB) at different levels of analysis. The CMM arose in NDB from the technical level and it provided the software projects at this organizational layer with prominent results. However, the CMM found serious limitations to progress in those areas at upper levels of management such as at the managerial and institutional layers. The organization NDB did not provide the necessary institutionalization of the CMM and nor the appropriate alignment of its goals with those of the CMM at the levels and areas which related the organization NDB to the market. Therefore, the organization NDB as a whole failed in the implementation of the CMM, but not the CMM itself.

Goals at the technical, managerial and institutional levels of the organization must satisfy (satisfice) criteria of alignment with each other in order to provide the organization with higher probabilities of survival and sustainable development. What makes the organization distinct in the achievement of its goals is the selection of its social structure (including processes such as the CMM), participants (decision-makers) and technology. If intelligence represents the ability of the organization to achieve its goals, then it is reasonable to assert that the higher the degree of organization cognition, the higher is the probability of the organization to satisfy its goals.

PART IV: GENERAL CONCLUSIONS

Never discourage anyone...who continually makes progress, no matter how slow.
Plato (427 BC – 347 BC)

Part VI provides general conclusions about this research and it consists of Chapter 6 only.

Chapter 6 starts by presenting a background about the theories and researchers who most influenced this research. It continues by highlighting its contributions and by discussing the alignment of its proposal with findings. Topics of further research are then pointed out and the chapter concludes by presenting perspectives about the implications of *cognitive machines* for organizations.

CHAPTER 6. BACKGROUND, CONTRIBUTIONS, EXTENSIONS AND IMPLICATIONS

6.1. Background to this Research

6.1.1. Simon and Zadeh's Theories

This research was influenced by the scientific work of Herbert A. Simon and Lotfi A. Zadeh developed in the period between the middle and the end of the 20th century. Despite having no direct relation to each other, they provided the literature with theories which under the perspective of this research complement each other by contributing important results to the fields of artificial intelligence, cognition, organizations and systems theory[42]. Such a background may be regarded as the first contribution of this investigation – i.e. to connect theories and results of these two brilliant researchers.

6.1.2. Bounded Rationality Theory

Simon was awarded in 1978 with the Nobel Prize in Economics. He received his PhD in Political Science from The University of Chicago in 1942 and among his prominent scientific contributions is the theory of administrative behaviour which comprises the concept of bounded rationality [Simon, 1982a, 1982b, 1997a and 1997b].

The theory of bounded rationality as proposed by Simon represents an important framework for the analysis of human behaviour, cognition and decision processes in organizations. It can also be viewed as a model of cognition and economic decision-making processes which considers the limits of knowledge and computational capacity of humans. However, Simon's theory of bounded rationality was missing alternative mathematical and computational tools which could be used to encapsulate the particularities of his model of human cognition and decision processes in a proper way. This was an important requirement for the development of the field of artificial intelligence – i.e. the need of alternative mathematical and computational approaches for the analysis, design and engineering of systems (machines) whose processes and behaviour are metaphors for, and models of, human cognition and intelligence.

Despite having important advancements since its inception in the early fifties, artificial intelligence has found serious limitations to progress in those areas where problems require approximate (fuzzy) rather than precise (crisp) formulation [Zadeh, 2001]. Such areas need alternative methodologies for the representation and manipulation of natural concepts which

---

[42] Simon and Zadeh played a counterpart task in the literature by proclaiming the lack of qualitative and quantitative approaches for coping with complex problems (where human behaviour, emotions and cognition are key factors). In his theory of bounded rationality, Simon called for new approaches which could extend the methods of decision analysis used by economists to a more realistic scenario on human decision-making [Simon, 1982b and 1997a]. In his work about systems theory, Zadeh pointed out the need for new mathematics in order to narrow the gap of understanding between the analysis of non-living and living systems [Zadeh, 1962]. This book advocates that such a new approach (as proclaimed by bounded rationality and general systems theorists) emerged with the advent of fuzzy systems theory [Zadeh, 1965 and 1973] and its derivatives on computing with words and perceptions [Zadeh, 1996a, 1999 and 2001] along with soft computing [Zadeh, 1994 and 1997].

are characterized by fuzzy boundaries [Nobre and Simon, 2003a].

6.1.3. Fuzzy Systems Theory

Zadeh received his PhD in Electrical Engineering from The University of Columbia in 1949 and among his prominent scientific contributions are the theories of fuzzy systems [Zadeh, 1965 and 1973], computing with words [Zadeh, 1996a] and computation of perceptions [Zadeh, 1999].

The theory of fuzzy systems represents an important framework with mathematical and computational background for the analysis of complex systems and decision processes – where complex systems is synonymous with systems (such as organizations) whose behaviour is preponderantly influenced by human emotion, cognition and social networks. The theories of computing with words and computation of perceptions are derivations of fuzzy systems and they represent approaches with the necessary elements to encapsulate the particularities of the Simon's model of bounded rationality. These particularities are mainly concerned with limitation of knowledge and computational capacity.

6.1.4. Connection between the Theories

Limitation of knowledge is synonymous with lack of information and also with the kind of uncertainty which pervades most of the concepts manipulated by humans. These concepts are called natural concepts and they are characterized by fuzzy boundaries [Bernstein *et al*, 1997]. Moreover, natural concepts form relations in propositions and clusters of propositions form mental models. In such a way, Zadeh's theories provide the necessary mathematical and computational background for the representation of natural concepts and mental models through complex symbols described by words and sentences of natural language.

Limitation of computational capacity is synonymous with the bounded ability of the human brain to resolve details and to solve problems with constraints such as time and cost. Such a limitation requires from humans the search for approximate solutions and satisfactory results rather than precise and optimal outcomes. In such a way, Zadeh's theories provide appropriate elements of approximate reasoning and economic decision-making which are necessary for the manipulation of natural concepts and mental models.

6.1.5. Results of the Connection

This work played an important part as a bridge between these theories of these two researchers. It put separate pieces of these theories together in order to:

- Bring the discipline of fuzzy systems and its derivatives (computing with words and computation of perceptions) closer to cognition, resulting in the design of *cognitive machines*.

- Relate *cognitive machines* with organizations and introduce them in the organization as participants and decision-makers.

- Use *cognitive machines* in conflict resolution (i.e. to solve intra-individual and group dysfunctional conflicts).

6.2. Research Contributions

6.2.1. On the Design of *Cognitive Machines*

In Chapter 2, it was defined that:

Definition 2.4.5.1: *Cognitive machines* are agents whose processes of functioning are mainly inspired by human cognition. Therefore, they have great possibilities to present intelligent behaviour.

In such a connection, this research played an important part by selecting technologies (which includes knowledge) of machines and by relating them to the discipline of cognition. The criteria used in the selection of the disciplines of fuzzy systems, computing with words and computation of perceptions were presented in Chapter 3 along with the design of *cognitive machines*.

In his work on a theory of computation of perceptions (which is a derivative of fuzzy systems), Zadeh introduced concepts which connect his theory to the process of perception [Zadeh, 1999 and 2001]. This research extended such connections by associating his concepts to additional elements of cognition such as natural concepts and levels of information-processing.

From such a background, Chapter 3 presented a framework of *cognitive machines* with the ability to manipulate complex symbols which are representations of percepts (and thus concepts) along with mental models described by words, propositions and sentences of natural language. The ability of these machines to manipulate natural concepts provides them with higher levels of information-processing than other symbolic-processing machines; and according to the theory of levels of processing in cognition [Reed, 1988], these machines mimic (even through simple models) cognitive processes of humans.

6.2.2. On *Cognitive Machines* and Organization Cognition

This research relied on the premise that *cognitive machines* can improve the cognitive abilities of the organization.

According to the literature, the participants within the organization are the main agents of organizational learning [Dierkes *et al*, 2001]. Chapter 2 introduced a similar perspective to the concept of organization cognition. It asserted that the participants within the organization are the main agents of organization cognition[43]. Additionally, in such a perspective the participants within the organization comprise humans and *cognitive machines* and they are supposed to act as decision-makers in the name of the organization. Therefore, *cognitive machines* are also agents of organizational learning and organization cognition.

6.2.3. On *Cognitive Machines* in Conflict Resolution: Analysis

---

[43] The organization is viewed as a cognitive system whose cognitive processes are attributes of the participants in the organization and the relationships or social networks which they form. These cognitive processes are supported by the goals, technology and social structure of the organization. Moreover, organization cognition is also influenced by inter-organizational processes and thus by the environment.

The premise that *cognitive machines* can improve the cognitive abilities of the organization was associated with the assertion that such machines can reduce or solve intra-individual and group dysfunctional conflicts which arise from decision-making processes in organizations. This assertion was supported by the analysis introduced in Chapter 4 through theories of bounded rationality, economic decision-making and conflict resolution along with perspectives about the participation of *cognitive machines* in organizations.

It was proposed that *cognitive machines* can reduce intra-individual conflicts by solving the problems of uncertainty, incomparability and unacceptability which pervades alternatives and the processes of decision-making of the participants in the organization. It was also proposed that such machines can reduce group conflicts by integrating the different views or opinions of the participants in a group within the organization into a commonsensical knowledge base and thus by making a commonsensical decision[44].

6.2.4. On Organization Cognition and Environmental Complexity

This research borrowed the picture of organizations as contingent upon the environment from the perspective of organization design and contingency theory proposed by Galbraith [1973, 1977 and 2002]. Moving further, it also relied on the proposition that an increase in organization cognition reduces the relative levels of uncertainty and complexity of the environment with which the organization relates. Such a proposition and the previous premise were summarized in one theorem as discussed in the following.

Theorem 2.5.1.1: The technology of *cognitive machines* increases the level of complexity of the organization (and thus the degree of cognition of the organization), and it relatively reduces the level of environmental complexity (and uncertainty) that the organization confronts.

According to a theory of hierarchic complex systems proposed by Boulding [1956], systems grow in complexity as they move from frameworks to social systems. The main element which makes such systems distinct from each other in terms of complexity and behaviour is cognition. Therefore, in this research organization complexity was defined as synonymous with (and contingent upon) organization cognition; and environmental complexity was defined as synonymous with (and contingent upon) environmental uncertainty.

Moreover, it was defined that organization cognition is a matter of degree which is contingent upon organization design – i.e. the choice of the elements of the organization such as its goals, social structure, participants and technology. Therefore, organization cognition differs from human cognition since the latter is part of a natural process.

Additionally, it was defined that organization cognition is contingent upon the environment. Therefore, organizations assume different degrees of cognition when they operate in different environments.

---

[44] A commonsensical knowledge base consists of a set of rules (conditional statements), and its design involves the integration of propositions and mental models of the participants in a group into a common memory (storage device). It represents an attempt to satisfy (satisfice) the perspectives of a group and criteria of design. A commonsensical decision involves a process with access to a commonsensical knowledge base. It attempts to make choices and to provide outcomes which satisfy (satisfice) the opinions of a group and criteria of design.

6.2.5. On the Participation of *Cognitive Machines* in Organizations

It was assumed in Chapter 1 that, if the cognitive roles in organizations, as fulfilled by agents, have performance and outcomes which can be attributed to humans or machines, without any distinction, then machines can be considered as participants within organizations.

Besides presenting the participation of *cognitive machines* in conflict resolution of decision-making processes in organizations, Chapter 4 proposed initial concepts about the relationships between the *cognitive machine*, its designer and the organization. It was defined that:

Definition 4.4.3.2: The organization is responsible for the assignment of roles to the *cognitive machine*, and the machine is responsible for the roles it serves the organization. However, the machine designer and the organization are the main parts responsible for the machine results and performance. If the machine presents deviant behaviour during task execution or performance below specified criteria, then the contract between the organization and the machine designer is the object of analysis and judgement.

6.2.6. On the Industrial Case

In Chapter 5 was presented an industrial case whose results contributed to support the premises, definitions and propositions of this work. The organization of study was NEC do Brasil S.A. (NDB) which is a subsidiary of the Japanese enterprise NEC (www.nec.com).

The first purpose of the industrial case was concerned with the implementation of the Capability Maturity Model (CMM) of the Software Engineering Institute of the Carnegie Mellon University in the organization NDB. Such an application involved practices of software process improvement and organizational learning. Its second purpose comprised the analysis and design of a *cognitive machine* and results about its participation in the adaptive learning cycle of the Radio Engineering Department of NDB. This application involved tasks of analysis, decision and management control of the performance[45] of large-scale software projects of the organization NDB.

- *On Organization Cognition, Process Maturity, Performance and Learning*

The results of the industrial case which were concerned with the process change management in NDB indicated that:

(i)   Organization cognition can be associated with organization process maturity. Hence, the degree of organization cognition can be represented by the level of process maturity of the organization.

(ii) The level of organization performance can be associated with the level of process maturity of the organization. Therefore, the degree of organization cognition can also be associated with the level of organization performance.

(iii) Improvements in organization cognition can be associated with improvement in organizational learning. This assertion is reinforced in the literature where improvements in organization performance and productivity are associated with the

---

[45] Performance was defined through tangible and intangible outcomes such as indexes of productivity and quality.

practices of organizational learning [Argote, 1999; and Brynjolfsson and Hitt, 2000].

- *On the Participation of a Cognitive Machine in the Organization NDB*

The results of the industrial case which were concerned with the technology change management in NDB indicated that:

(i)   *Cognitive machines* can support managers in the analysis of decisions and management processes of the organization performance. Such tasks involve a combination of manipulation of natural concepts with arithmetic and analytical calculation.

(ii)   The engineering of such machines involves the design of a commonsensical knowledge base which integrates different perspectives and perceptions of the participants in a group. Additionally, the ability of these machines to represent natural concepts and to manipulate mental models through rules of approximate reasoning makes them candidates to reduce intra-individual and group dysfunctional conflicts which arise from decision-making processes.

- *On the Analysis of the Technical, Managerial and Institutional Levels of NDB*

Observations about the organization levels of analysis of NDB indicated that:

(i)   The Capability Maturity Model (CMM) provided the technical level of NDB with successful results which could be observed through improvements in the levels of organization process maturity and performance of software projects.

(ii)   The CMM found limitations to progress in those areas at upper levels of management such as at the managerial and institutional layers. It was caused by the lack of alignment between the organization goals with those of the CMM at the levels and areas which related the organization NDB to the market. Therefore, the organization NDB as a whole failed in the implementation of the CMM, but not the CMM itself.

- *On Organization Cognition and the Environmental Complexity of NDB*

It was also observed that during the period between 1995 and 1998 the Brazilian telecommunications market, and thus the environment of NDB, passed through a drastic transformation as a result of a constitutional process of privatization. Between 1998 and 2002, the number of participants (employees) of NDB was reduced by 90% approximately.

These additional observations about NDB lead to the conclusions that:

(i)   The alignment of the technical, managerial and institutional levels of the organization represents a necessary process for its survival and sustainable development. Such a process involves the alignment of goals and social structure.

(ii)   Organization cognition is an organizational ability[46]. Therefore, organizations presenting a satisfactory degree of cognition in only isolated parts or hierarchy levels may fail to survive and to develop.

---

[46] Agents of organization cognition include the participants in the organization, the relationships which they form, inter-organizational processes and environmental factors.

6.3. Further Extensions

6.3.1. On *Cognitive Machines* and Learning

The design of a framework of *cognitive machines* introduced in Chapter 3 was mainly concerned with decision-making processes and the topic on learning was left for further research.

The field of machine learning is concerned with the engineering of computational systems that automatically change and improve with experience [Mitchell, 1997]. Neural computation [Hertz, 1991], soft computing [Zadeh, 1994], adaptive fuzzy systems [Wang, 1994], evolutionary computation and genetic algorithms [Back, *et al*, 2000; and Fogel, 2000], along with genetic programming [Koza, J.R. 1992] are disciplines which can aggregate processes of learning to machines.

The processes and algorithms of learning introduced by Wang [1994] represent potential candidates for such a purpose – i.e. to provide the *cognitive machines* of this research with the ability to learn. The framework of *cognitive machines* introduced in Chapter 3 has structure and processes similar to those of the fuzzy systems adopted by Wang. Therefore, similar algorithms of learning could be used.

6.3.2. On *Cognitive Machines* and Emotions

The topic of machines with emotions and emotional processes was also left for further research. However, it deserves some comments due to its importance in the literature.

Whether machines should exhibit emotional behaviour, and whether they are able to have emotions or not, are controversial topics within artificial intelligence, cognition and the social sciences.

By assuming that machines may indeed be able to have emotional processes and behaviour, the question of whether emotions are important to machines or not depends on the purpose of its designer and the environment with which they will relate. On the one hand, machines with emotions might form better relations and social networks with humans in organizations than other machines. In such a view, machine emotion would be relevant for organizational behaviour theorists. On the other hand, machines with emotions might have their own motives and might represent additional agents of dysfunctional conflicts in organizations. In such a view, machine emotion would be a problem for rational theorists.

Among the Institutions worldwide researching the field of machine emotion include The MIT Artificial Intelligence Laboratory at Massachusetts (USA) which has carried out a project called Sociable Machines [Breazeal, 2000].

6.3.3. On *Cognitive Machines vs.* Humans in Organizations

Are *cognitive machines* better agents of organization cognition and organizational learning than humans? Are they better agents of organization performance and productivity than humans? Such questions rely on the statement that: if one assumes that the cognitive roles in organizations have performance and outcomes which can be attributed to either humans or machines, without any distinction, then one is ready to consider machines as participants within the organization similarly to people. This perspective involves a rational comparison of machines with human performance and thus they compete for the same role in the organization.

Another formulation for such a problem consists of asking whether *cognitive machines* can satisfy (satisfice) cognitive roles and goals in the organization or not. In this perspective, machines are not competing with humans directly, but with the organization criteria of satisfying (satisficing) roles and goals. Both perspectives need to be further investigated in order to present economic criteria of choice and analysis of social implications.

6.4. Implications of *Cognitive Machines*

Imagination is more important than knowledge...

Albert Einstein (1879-1955)

To predict the implications of *cognitive machines* for organizations is a topic of further research which demands a high level of imagination in connection with the knowledge available in the literature. Under such a perspective, this section presents a brief overview about some of the possible implications. It starts by discussing literature on the benefits of information technology (IT) to organizations. It proceeds by connecting IT with *cognitive machines*, and thus by presenting the implications of the latter for organizations.

6.4.1. On Information Technology and Economic Growth of Organizations

Investments in information technology (IT), which involves computers, software programs and communication systems, have demonstrated to be connected with improvements in tangible and intangible outcomes of organizations such as economic performance, human capital and quality of services and goods. Results of case studies at the firm level of analysis have shown that the impact of IT on economic growth of organizations is relevant and quite large when compared to their share of capital stock or investment, and this impact is likely to grow more in the next few years [Brynjolfsson and Hitt, 2000].

Improvements in organizations due to investments in IT are associated with:

(a) Reduction of costs of coordination, communications and information-processing. It is quite reasonable to assert that nowadays managers make decisions which involve much more inter-dependent variables than in the pre-computer era [Simon, 1977 and 1982b]. In a metaphorical way, computers play the role of manipulation of variables behind the curtain of managerial complex decisions.

(b) Complementary investments in organizational processes and human capital which also lead the organization to improve intangible outcomes such as quality [Brynjolfsson and Hitt, 2000].

6.4.2. On Information Technology and *Cognitive Machines*

Cognitive machines are special classes of information technology (IT) and thus they represent potential candidates which can provide organizations with additional improvements in those areas where IT has already demonstrated its strengths.

6.4.3. On *Cognitive Machines* in the New Organization

- ***Herbert A. Simon's Statement on Computers in Organizations***

> *In our fascination with change, we must keep firmly in mind that the structures of effective organizations are at least as much shaped by the tasks they are designed to perform as by the nature of the human and computer resources available for performing them. The permanence of most of these tasks constitutes the most important reason for predicting that the organizations of the future will, in most respects, resemble the organizations of the past and present. They will retain the familiar shape of a hierarchy of semi-independent components and sub-components.*
>
> *Herbert A. Simon (1916-2001), published in [Augier and March, 2002].*

This research supports such a statement and it extends the term "computer resources" to include *cognitive machines*. Moreover, it discusses in the following some implications that *cognitive machines* may have on organization design and its elements.

- ***Organization Design***

The perspective of organization design as a process of choice of organization models will not change in the future. The strategic selection of the goals, social structure, participants and technology of the organization will remain contingent upon the environment with which it relates. The characteristics of the elements of the organization will change, evolve and develop continuously, in complexity and knowledge, but the purpose of existence of the organization will remain the same or will not change in the same proportion of its elements.

- ***Goals***

Might *cognitive machines* be able to have goals or motives? The answer would be yes if they would be viewed as carriers of people's motives and organization goals. They might be able to use their own strategies and tactics as means to satisfy (satisfice) organization goals and to support people in the achievement of their own motives[47].

- ***Social Structure***

A social structure comprises normative and behavioural parts as defined in Appendix B. While a normative structure consists of institutionalized rational procedures and prescriptions for behaviour, a behavioural structure consists of actual behaviour which emerges from non-procedural processes and social relations.

The introduction of *cognitive machines* in organizations, with no (or little) attention to emotions, will provide the organization with higher levels of rationalization since its normative structure will overcome non-procedural behaviour. This phenomenon can lead the organization to higher levels of predictability of outcomes which depend on prescriptive behaviour.

---

[47] People's motives in organizations involve consciousness. "*To be* conscious that we are perceiving or thinking is to be conscious of our own existence*": Aristotle (384BC – 322BC).

- *Participants*

According to the Cambridge International Dictionary of English (Cambridge University Press, 2000), a participant is "a person who takes part in or becomes involved in a particular activity".

By assuming the perspective that computer systems take part in and are involved in activities of information-processing and communications in organizations of today, then it is reasonable to assert that *cognitive machines*, like people, are participants within the organization.

However, by considering that participation in the organization involves consciousness of the participant; one might question whether *cognitive machines* are able to be conscious. This is another controversial subject in the fields of artificial intelligence, cognition and social sciences which demands further investigation. Nevertheless, it was defined in Chapter 4 that:

Definition 4.4.2.1: Machine consciousness represents the awareness of its designer in relation to the cognitive processes and abilities that the machine carries on during task execution.

- *Inducements (Rewards and incentives) and Contracts*

From the perspective that *cognitive machines* are carriers of people's motives and desires, it is plausible to assume that they do not need to be induced by the organization directly, but the designers of these machines are who do need receive incentives. Therefore, a contract must exist between the organization and the machine designer.

- *Technology*

Cognitive machines might demand investments in alternative business processes and technologies from the organization. Such investments can represent important advantages to improve tangible and intangible outcomes of the organization such as demonstrated in the literature of economic perspectives of information technology in organizations [Brynjolfsson and Hitt, 2000].

- *Cognition as an Agent of Organization Change*

While the structure or anatomy of the organization of the future might remain similar to the models of the past and today, its processes or physiology might change in greater proportions. Advancements in the science of cognition will continue to represent an important agent of organization change.

Developments in cognition provide researchers with a better understanding of the organization of the human brain and its processes, and discoveries in this field contribute to extend other disciplines to new frontiers. In the area of organizations, for instance, researchers will be able to provide the organization with alternative processes of learning and decision-making, and thus with more complex models of organizational learning and organization cognition. Advancements in cognition will also provide *cognitive machines* with more realistic models of the human mind, and thus with the ability to solve more complex problems which involve natural or fuzzy concepts. Moreover, education might be the area of foremost transformation as more developments in cognition are reached.

- *The New Organization as Telecommunications Management Networks*

The idea of the new organization resembling Telecommunications Management Networks [ITU-T, 2000] was touched upon in [Nobre and Simon, 2001b and 2002]. It assumes a high level of automation for the new organization, where machines will perform most of its technical and managerial activities. Viewed as information-processing elements, these machines will need to interact with each other via standardized protocols of communication systems. Such interactions will need to follow processes and to be organized into architectures of hierarchic layers.

APPENDIX A. ANALYSIS AND DESIGN OF THEORY

A.1. Nature of Theory

Developing a theory is a scientific enterprise. It is designed to explain phenomena. If the phenomena can be explained by a theory, then the theory can facilitate predictions about the future behaviour of the phenomena. The theory gets confirmed once explanations and predictions are borne out, and this gives the scientist greater control over the domain of the phenomena [Khandwalla, 1977].

Therefore, first of all theories consist of a formal background that explains past and current phenomena with a subject of study. Secondly, in addition to explanations, a theory has also to provide knowledge to support predictions. Thirdly, theories have to provide evidence in order to support their explanations and predictions.

Explanatory studies attempt to answer why and how things happened, but not only to tell what happened as descriptive studies do. Explanation and prediction are complementary processes. The former takes place after the event has occurred, and the latter extends the former to the control of events that may occur.

A.2. Components of Theory

The basic components of theory are variables, propositions and reasoning processes. The former can assume concepts of mental images or perceptions, words of a natural language, numbers and general symbols. The second is simply a statement about one or more variables and it defines the relationships between the variables of study [Bailey, 1982]. The latter can be synonymous of logic and cognitive mechanisms used to deduct and to induct new propositions from either previous propositions or qualitative and quantitative data analysis.

Propositions encompass hypothesis, axioms, postulates and theorems. Hypotheses are often defined alone and they can be used to induct new hypotheses by either generalisation or analogy. On the other hand, axioms and postulates are provided in the form of a set of statements whose combination can be used to deduct through syllogism additional propositions called theorems.

A.2.1. On Hypothesis and Inductive Reasoning

A hypothesis is a proposition that is stated in testable form and predicts a particular relationship between two or more variables.

Inductive reasoning works basically in two ways. Firstly, it moves from specific observations, qualitative and quantitative data, statistical analysis results and thus hypotheses, towards broader results and theories similarly to a bottom-up approach. The generalization of a hypothesis is an attempt to assert that it holds in all cases, or in most of the cases under investigation. Hence, this process is also called induction by enumeration, or, more commonly, generalisation. Secondly, inductive reasoning can also draw conclusions and hypotheses about things based on their similarities to other things, and thus this process is called induction by analogy.

Figure A.1 illustrates the primary steps of an inductive process. It begins with specific observations, proceeds with data analysis to detect patterns and regularities, formulates some tentative hypotheses that can be explored, and finally ends up developing some general conclusions, hypotheses and theories. A horizontal way is also drawn to show the process of induction by analogy.

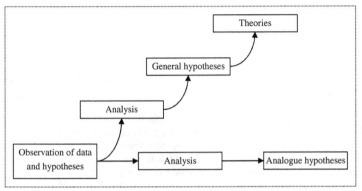

Figure A.1. Inductive Processes of Generalisation and Analogy

A.2.2. On Axioms, Postulates, Theorems and Deductive Reasoning

On the other hand, deductive reasoning works from the more general to the more specific, similarly to a top-down approach. One might begin with a theory and propositions about a specific topic. One then may narrow them down into more specific propositions called theorems that one can test. One narrows down even further when one collects observations and data to address the theorems. This ultimately leads one to be able to test the theorems with specific data, providing evidence to the original theory. Figure A.2 illustrates such a process of deductive reasoning.

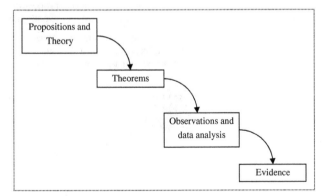

Figure A.2. Deductive Process

Axioms, postulates and theorems are components of axiomatic or deductive theory, which takes the form of a set of interrelated propositions as in the following example.

Proposition 1: If A then B

Proposition 2: If B then C

Therefore,

Proposition 3: If A then C

In such a theory, if propositions 1 and 2 are true statements, it follows by deduction that proposition 3 is also true. Such true statements are called axioms and postulates, and other statements can be deduced from them. Thus propositions 1 and 2 of the previous example are axioms and postulates. Nevertheless, axiom has a mathematical connotation and is used more often for statements that are true by definition and for propositions involving highly abstract concepts. Postulate is more often used for statements whose truth has been demonstrated empirically. Additionally, a proposition that can be deduced from a set of statements is called theorem. Proposition 3 in the previous example symbolises a theorem deduced from the statements 1 and 2. As an example, consider the propositions settled in the following where postulates 1 and 2 are used to derive theorem 1 by deductive syllogism.

Postulate 1: The better the incentives and reward systems, the more productive the workers of an organization are.

Postulate 2: The more productive the workers are, the higher is the probability of an organization to satisfy its goals.

Therefore,

Theorem 1: The better the incentives and reward systems, the higher is the probability of an organization to satisfy its goals.

Since postulates are considered to be true, there is little reason to treat them as testable hypothesis. However, it makes often necessary to write the deduced proposition - i.e. the theorem - as a hypothesis and test it, as this is the main mean of testing the entire theory.

Despite distinct, inductive and deductive processes complement each other. The former can be used to the exploration of new theories and broader results. The latter can be used to the exploitation of propositions and theories, providing them with refinements and more precise results.

Table A.1 summarises the different types of propositions and their scope of application.

Table A.1. Types of Propositions

Propositions	Derivation	Testable
Hypothesis	Induced by generalization or analogy	Yes
Axiom	True by definition	No
Postulate	Assumed to be true	No
Theorem	Deduced from axioms or postulates	Yes

A.3. A Process of Theorizing

Theorizing is a process which involves the explanation and prediction of natural and artificial phenomena, and it encompasses evidence [Bailey, 1982].

In order to formulate a theory, firstly one needs to choose a problem among alternatives in order to proceed with the analysis of its domain. Secondly, after analysing the problem, one needs to select the set of variables that represent the domain of interest. Moreover, one needs to decide how to measure such variables and also to design propositions that define the relationships among them. Thirdly, one needs to gather data about the variables from a sample. Fourthly, after gathering the data one needs to analyse them. Moreover, one has to add meaning to them and to the relationships among them. Fifthly and lastly, one needs to provide evidence which supports the theory. It is done by testing propositions and by analysing the results.

Such a process can provide a theory with better results as one circulates from the first to the fifth stages continuously. At the end of step 5 one may needs to analyse the problem again, to redesign its variables and propositions, to refine the gathered data and to analyse them again, and finally to review the evidence and results. Figure A.1 depicts such a continuous process.

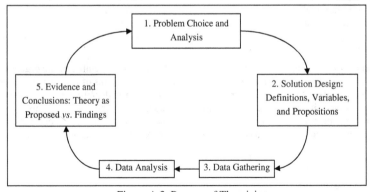

Figure A.3. Process of Theorizing

A.4. Approaches to Organization Studies

This section presents some of the main methods for studying organizations, and in particular, those methods which provide a theory on organizations with data and evidence [Khandwalla, 1977]. For the purpose of this research, such a set of methods will be also called "approaches to organization studies".

A.4.1. Case Studies

The researcher interviews a few individuals to determine the organization background and history. The interview can include questions on: age, size, structure and wealth of the organization; number of employees, markets of action, goals, strategies, strikes and conflicts; levels of centralization and decentralization; specialization; the technology employed by the organization; authority, power and responsibilities with the organization; reward systems; the processes with the organization; the relationship between the organization and the environment; and so on.

Case studies are likely to yield incomplete information because of the few individuals interviewed and also due to it involves the study of only one or a very few organizations.

A.4.2. Field Studies

Shortly speaking, field studies and case studies are alike. However, field studies are usually much more time consuming than case studies since the researcher tries to interview a more significant sample of organization members.

A.4.3. Participant Observation

This method, and the two previous ones, can be classified within the broad spectrum of field approaches [Scott, 1965]. The researcher is a member of the organization, and being a participant, he derives data and conclusions by observing facts within the organization behaviour.

The researcher can join the organization with the purpose of observing its behaviour, like a consultant for instance. Nevertheless, despite his comfortable position to observe how those around him really feel and what really go on within the organization, this method also results in serious drawbacks. People within the organization can feel constrained with the researcher's presence, and emotional involvement with them may also give the researcher certain biases.

On the other hand, if the researcher is a current or previous member of the organization, thus he can provide a formal study on it without any of the problems of feelings as mentioned before. Nevertheless, his experience may be limited to a specific department or division of the organization and it should be considered when addressing and reporting his research. Therefore, the more access and participation within the different departments and activities of the organization, the broader and richer can be the researcher's investigation and conclusions.

A.4.4. Questionnaire Survey

It consists of gathering information from the organization (or from a sample of organizations) by interviewing its members and by requesting their answers to questionnaire surveys. If quantitative data are provided with the surveys, thus this method makes possible the development of statistical analysis. This method can also be used to support the previous approaches.

A.4.5. Field Experiment

In this approach, the researcher tries to manipulate some variables in order to observe their effects on other variables which affect the functioning of the organization. Field experiments are not appropriate to organizations since they can put them in situations of risk. Nevertheless, such a method can be used in laboratories in order to simulate the behaviour of the organization.

A.4.6. Literature Review

It consists of a review of published material on the subjects of organizations which matches the interest of the researcher - such as decision-making and problem solving, the exercise of power and authority, and so on. March and Scott provide a distinguished literature review on organizations, including various levels of organization analysis [March, 1965; and Scott, 1998]. Literature review is an important method which supports the researcher in the proposal of new theories on organizations.

A.4.7. Information Survey

It consists of gathering information about a representative number of organizations as provided by the literature - like by The Journal of Economic Perspectives and The American Economic Review for instance - but also by consistent sources of data - like The United Nations and World Bank for instance. Information survey is an important approach to provide researchers with qualitative and quantitative data which can be used to justify theories, generalize results and also to derive new propositions on theories.

Information survey differs from literature review since it is concerned with data collection about organizations and markets, including return on investment and gross domestic income of a country for instance. On the other hand literature review is more concerned with the study of principles and theories of organizations.

A.4.8. Analytical Research

It consists of building mathematical models of organizations and it has been a common approach used by economists to the analysis and design of processes within organizations and between networks of organizations [Shy, 2001]. Such an approach provides reductionism by the nature of mathematics, since the problem under analysis is usually formulated by considering a set of constrains [Helm, 2000]. However, it brings clarification and accuracy to the domain of the problem under analysis [Simon, 1957; and Starbuck, 1965].

Nevertheless, organizations hold properties of dynamic and non-linear systems, coupled with the environment, with capabilities to adapt and to evolve, and whose behaviour results from interactions among a variety of adaptive agents and other factors. Therefore, organizations seem to be poor candidates for analytical approaches. Such behaviour characterizes organizations as complex systems whose degree of complexity can only be poorly modelled by analytical approaches [Prietula *et al*, 1998].

A.4.9. Computational Modelling and Simulation

The approach to computational modelling and simulation of organizations has received new insights mainly from the 1990's [Carley and Gasser, 1999], and thus special attention is devoted to this subject in the following.

Models are simplified representations of objects and phenomena encountered within the world. Therefore, as analytical models are, computational models are also characterized by reductionism, since they include constrains and approximations of the problem under analysis. Nevertheless, computational modelling has developed as an alternative and powerful tool to support the analysis and design of systems of higher magnitude of complexity than those treated by analytical methods.

The capability of computational models to overcome the limitations of analytical approaches in the modelling of complex behaviour in organizations is only one of its main advantages. Compared with experiments using human subjects, computational models are generally less noisy, easier to control, more flexible, and can be used to the analysis of a larger variety of factors within less time. Furthermore, computational models may be larger, to include more agents, and may cover a longer period with more tasks than can be covered in a human laboratory experiment.

In the same way as analytical modelling, computational modelling tries to achieve more precise and well-defined results for organizations than theories which involve vague and intuitive propositions. Additionally, computational modelling and simulation can be seen as a hypothesis generator system with capabilities to derive propositions and to verify their consistency to theoretical conclusions. Hence, theories can be confronted with organizations, and new theories can also be derived from simulations of computational models.

The pioneer applications of computational modelling to the analysis of organizations appeared in the second half of the 20th century [Cyert and March, 1963; and Cohen and Cyert, 1965]. Computational modelling of organizations advanced after the development of digital computers, and most importantly, it received new insights with the advent and maturation of the disciplines of artificial intelligence, cognitive science and information processing systems, social psychology and multi-agent systems. Perspectives on organizations resembling cognitive and information processing systems with computational agents of bounded rationality emerged during that period [Simon, 1947; and March and Simon, 1958], and thus they opened new doors to the research of new areas like computational organizational theory.

Computational organizational theory (COT) has emerged as a scientific base to study organizations as distributed computational agents [Prietula *et al*, 1998]. COT encompasses distributed artificial intelligence, agent technology, software engineering and organization theory as its principal disciplines, where the latter involves mainly the domains of organization behaviour, sociology, economy, psychology and political sciences.

Within the field of COT, computational analysis is used to develop a better understanding of the fundamental principles of organizing multiple information processing agents and the nature of organizations as computational agents. Research in this area has two focuses. The first has to do with building new concepts, theories and knowledge about organizing and organizations. The second has to do with developing tools and procedures for their validation.

A.5. Criteria of Choice of Research Methods

Each research method or approach to organization studies presented in the previous section has its advantages and disadvantages. Therefore, it makes necessary to define a set of criteria in order to support the researcher to choose one or more methods among the set of available alternatives. Table A.2 presents a set of criteria adapted from [Khandwalla, 1997], which are useful in the selection of approaches to the study of organizations.

Table A.2. Criteria of Choice of Research Methods

a. How much control does the researcher have on the manipulation of the variables?   (criteria of control)
b. How much can the results gotten through the use of the method be generalised to other organizations?   (criteria of generalisation)
c. How valid and reliable are the measurements of organization variables?   (criteria of reliability)
d. How economically can the researcher get the necessary information?   (criteria of economy)
e. How speedily can the researcher get the necessary information?   (criteria of speedy)
f. How much knowledge does the researcher have on the method?   (criteria of application)

Table A.3 presents the results of application of each criteria of Table A.2 to the research methods.

Table A.3. Application of Criteria to Research Methods

Research Method	control	generality	reliability	economy	speedy	application
Case Study	low	low	low	high	high	low
Field Study	low	low	high	medium	low	low
*Participant Observation	medium	low	high	high	high	high
Questionnaire Survey	low	high	medium	medium	medium	medium
Field Experiment	medium	low	high	low	low	low
*Literature Review	low	medium	medium	high	medium	high
*Information Survey	low	high	high	high	high	high
*Analytical Research	high	low	N.A.	high	medium	medium
*Computational Simulation	high	medium	N.A.	high	medium	medium

*Selected research methods to the study of organizations within this investigation.
Low, medium and high denote the magnitude of application of each criterion to the research methods.
N.A. means not applicable.

It must be considered that different researchers may have distinct opinions on the application of the criteria of Table A.2 to the set of research methods shown in Table A.3. Criteria (f) for instance, that concerns application, depends on the knowledge of the researcher about each method, and thus it plays an important role in the process of choice.

Table A.4 was designed to add quantitative meaning to the results of Table A.3 and to reduce their subjectiveness. For such a task, the values 1, 2 and 3 were attributed to the labels low, medium and high respectively, and thus a weight could be computed to each research method according to the average of such values.

$$w_i = \frac{\sum_{j=1}^{n} c_j}{j}, \text{ for } i = 1,\dots,9 \text{ and } j = 1,\dots,n \tag{A.1}$$

where w_i denotes the weight computed to each respective research method i; c_j represents the values attributed to each criteria; and n is the number of criteria applied to account the average. It must be observed that $n = 5$ for the methods of computational simulation and analytical research only, since N.A. cannot be accounted. On the other hand, $n = 6$ for all of the others.

Table A.4. Attribution of Weights to Research Methods

Res. Method	control	generality	reliability	economy	speedy	application	weight
Case Study	1	1	1	3	3	1	1.7
Field Study	1	1	3	2	1	1	1.5
*Part. Observ.	2	1	3	3	3	3	2.5
Quest. Survey	1	3	2	2	2	2	2.0
Field Exper.	2	1	3	1	1	1	1.8
*Literat. Rev.	1	2	2	3	2	3	2.2
*Info. Survey	1	3	3	3	3	3	2.7
*Analyt. Res.	3	1	N.A.	3	2	2	2.2
*Comp. Simul.	3	2	N.A.	3	2	2	2.4

*Selected research methods to the study of organizations within this investigation.
Low =1, medium =2 and high = 3.
N.A. is not accounted to the average.

Figure A.4 depicts a graphical representation of the weights computed to each research method. CS, FS, PO, QS, FE, LR, IS, AR, and CM abbreviates the respective research methods.

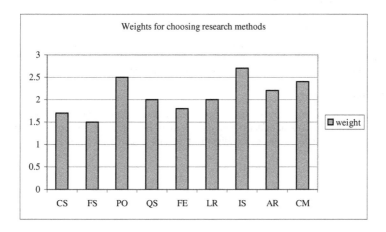

Figure A.4. Applied Criteria and Computed Weights to Research Methods

A.6. Selected Approaches to Organization Studies within this Research

It can be asserted that the higher the weight value of a research method, the more likely is it to be chosen among the set of alternatives. By observing Figure A.1, it can be stated that the approaches to participant observation (PO), literature review (LR), information survey (IS), analytical research (AR) and computational modelling and simulation (CM), are the stronger candidates for the study of organizations within this work. Hence, such methods were selected according to these criteria. Moreover, the analysis in the following provides additional rationales for these choices.

- *Literature Review*

Literature review is fundamental to support a new theory on organizations. After providing an overview on organizations, thus current principles and theories can be selected, unified and extended to form new concepts. That is the main point, i.e. to understand the past and the current trends on organizations in order to propose new ones.

- *Analytical Research*

Analytical research is used within this work when analysis needs to be reduced to accurate models of organizations. This approach is mainly used to support some of the definitions of organizations as exposed in Part II, and also to demonstrate the consistency of the analysis and design of a computational model as proposed in Part III.

- *Participant Observation*

Analyses of a distinguished industrial firm are provided in Part III since the author was a member of such a firm during the period between 1997 and 2000. Such an investigation is broad enough to conclude results to distinct levels of the firm, ranging from technical and managerial to institutional analysis.

- *Computational Modelling and Simulation*

Computational modelling and simulation is used to support the analyses provided by the approach to participant observation. Through this method, models can be built and tested, and variables can be manipulated in order to derive new conclusions on the whole analysis.

- *Information Survey*

Information survey provides a basis to support conclusions and to give evidence. It provides this work with a large amount of data about its domain of investigation - i.e. organizations and technology - in order to measure the alignment of this research proposal and data of the market.

APPENDIX B. ORGANIZATION THEORY

B.1. Introduction

This Appendix surveys organization theory and it puts such a multi-disciplinary field into the context of this research. It starts by presenting a historical perspective on the evolution and development of organizations towards organizations of today; it overviews modern schools of organization theory as developed during the 20th century; it presents complementary perspectives on organizations as rational, natural and open systems; it proposes some rationales for organizing into political, economic and social contexts; it explains the benefits of organizations; it presents a model of the organization and describes its elements; it defines organization theory as a discipline for the analysis and design of organizations; it explains the nature of organizations in order to justify the diversity of definitions about them in the literature; and it reviews the concept of formal organizations. Additionally, this Appendix is complemented by Appendix C which surveys the main disciplines of foundation of organization theory.

B.2. Organizations in History

The synthesis about the history of organizations presented in this section represents a survey about the literature, and therefore it is not an invention. Nevertheless, it contributes by unifying peaces of information gathered from selected references about: the history of Europe [Delouche, 2001]; the history of management thought [George, 1972; and Wren, 1987]; organizations [Scott, 1998]; and global analyses of economic and social benefits of organizations and technology [Easterlin, 2000; Gordon, 2000; and Johnson, 2000].

Organizations have gradually grown in importance within the economic, social and political worldwide contexts according to the evolution and development of the human history. They found their maturation after the Industrial Revolution originated in Europe in the 18th century and later spread in the United States of America in the 19th century. The gradual transition from a non-industrial to an industrial society has marked the frontiers between the periods of evolution and development of organizations.

The term evolution is usually used when changes in the society assume relatively unpredictable forms. Instead, development follows a more predictable sequence of stages deliberately planned with some form of modernisation as the intended end [Richter, 1982]. The former encompasses the processes of organizing provided by ancient and the Middle Ages civilizations. Nevertheless, the flourishing of the Renaissance in Europe emerged as a gear lever to support the latter, and thus the development of the Industrial Revolution in Europe.

B.2.1. Old Organizations: Ancient Civilizations and the Middle Ages

Many principles of today about organizing emerged during ancient civilizations (5.000 B.C. - 500 A.C) among Sumerians, Egyptians, Babylonians, Hebrews, Chinese, Greeks and Romans. It is highly probable that organizing processes first began in the family, later expanding to the tribe, and finally pervaded the formalized political units [Wren, 1987].

After the fall of the Roman Empire in A.D. 476 and the subsequent emergence of the Feudalism in Europe during the Middle Ages, new principles of organizing evolved as solutions to the economic and political crises in Europe. Although organised in a feudal structure, man began to take significant steps in his thinking about organization and management. An increasing record of writings about the discipline of organizing also characterized the Middle Ages. Nevertheless, economies and societies were essentially static, management practices were still largely antihuman, and political values involved unilateral decisions by some central authority. All of these conditions were not favourable to develop an industrialized society with some liberty and market ethic.

B.2.2. Pre-Industrial Organizations: The Renaissance and the Enlightenment Age

The crises in Europe during the 14th and 16th centuries paved a revolution in thinking, together with religious, social, economic and political strife, giving genesis to the period of Renaissance in the 16th century. The Renaissance brought a new focus on reason, discovery, exploration and science, and thus a revolution in scientific thought began in the 16th and 17th centuries. The overseas expansion of Europe between the 16th and 18th centuries strengthened the integration of cultures of different continents giving genesis to the Mercantilism. Such revolution with the strengthening of globalization and characterised by a world scale economy, requested the development of more complex principles of organizing.

Additionally, in the period of Renaissance there was an increasingly need and call for practices that could bring ethics to the liberty and to the market, forming the philosophy of the Enlightenment Age. At that Age, political philosophers began to stimulate the thoughts of people by disseminating new ideas about equality, justice, the rights of citizens, notions of a republic governed by the consent of the people, and to deliberate concepts such as decentralization rather then centralization of power.

In the 18th century new economic theories emerged to challenge Mercantilism and the controlling power of the landed aristocracy. In his work on *Wealth of Nations*, Adam Smith (1723-1790) established the classical school of liberal economics and he proposed that only the market and competition would be the regulators of economic activity. The Enlightenment Age had then opened the doors towards a new era called Industrial Revolution.

B.2.3. Modern Organizations: Post-Industrial Revolution

The transition from pre to post-industrial organizations was not sharp, and instead, it happened gradually. Such a transition created new social, economic and political conditions and it brought new challenges to the society as a whole. The continuous advances in science and technology made possible large combinations of humans and machines, giving the genesis for a new generation of organizations. Therefore, principles of organization and management had to be revised, improved and extended to a new cultural environment.

The post-industrial generation of organizations was characterised by the agglomeration of people into a central place, and also by the introduction of machines to minimize the work of human's muscles. Such organizations received new insights on organization design - like those of division of labour, departmentalization, centralization and decentralization of decisions, incentive and reward systems, and so on. Scientific management thoughts were developed at the same time in order to integrate functions of planning, leading, coordinating and controlling to the whole organization [George, 1968; and Wren, 1989].

Therefore, organizations gradually started to liberty people from the primitive use of muscles as an essential condition for survival. Machines provided people with economy in time and strength, and thus extra efforts could be devoted to the exercise and exploration of more complex cognitive tasks in organizations. However, modern organizations increasingly started to depend on the use of additional resources such as energy in order to feed their electrical, combustion and steam engines. Moreover, until late of 1940's, little attention was given to the organization's environment.

B.3. Organizations of Today

Similarly, the transition from post-industrial to the organizations of today has been defined gradually.

Organizations have been shaped by advances in computers, communication networks and general information technologies. Moreover, they also have been influenced by cultural, normative and regulative institutional processes.

Tracing back to the 18th, 19th and 20th centuries, organizations (of manufacturing) have gradually replaced muscular activities of humans with machines. It was possible after the invention and commercialization of internal combustion engines, electricity and electric motors. Proceeding further, and now from the second half of 20th century, organizations have replaced some cognitive abilities of humans with new machines. The use of these machines to support and to carry out mental tasks in organizations of today has given people additional resources to expand the frontiers of the organization. Economy of time and energy (both physical and mental) has led people to overlook the organization at upper levels of analysis. Hence, they have developed new processes for the peripheral components of the organization, and in particular to the levels that connect the organization to the environment.

As in the post-industrial era, organizations of today still demand, and depend on, energy-based technologies. However, with advancements in information-processing technologies, the demand for information has become relatively greater than energy. Therefore, organizations have continuously shifted attention from energy to information.

The gradual transition from energy to information management has motivated researchers to develop theories and perspectives on organizations as cognitive and learning systems [Carley and Gasser, 1999; Dierkes *et al*, 2001; March and Simon, 1958]. In these perspectives, the environment is viewed as a dynamic and complex system that influences the organization (and vice-versa).

B.4. Schools of Organizations: The 20th Century

Theories of organizations have been properly developed since the beginning of the 20th century. Their initial ground was mainly prepared from previous theories in philosophy and social sciences. However, organization theorists simultaneously advanced in knowledge with the development of new theories and sciences. This work asserts that organization theory has matured with the chronological developments in management, psychology research and general systems theory.

Organization theory constitutes a multi-disciplinary field and thus it has received contributions from scientists of diverse areas and different backgrounds. The literature has classified the major contributors of organization theory in different schools, according to their lines of investigation [Grusky and Miller, 1981; Pugh, 1997; and Scott, 1998].

The schools presented in his section follow a chronological order of development back through the 20th century. However, administrative behaviour (within decision-making and bounded rationality), general systems, contingency theory and organizational learning are the schools of organizations which play the most influential part in this work. Table B.4.1 names the schools as described in this section in order.

Table B.1. Major Schools of Organizations of the 20th Century

Year of Publication	Schools of Organizations (and Major Contributors)
1924	Bureaucracy (Max Weber)
1911	Scientific Management (Frederick Taylor)
1916	Administrative Theory (Henri Fayol)
1945	Human Relations (Elton Mayo)
1947 / 1958	Administrative Behaviour and Decision-Making (Herbert Simon and James March)
1966 /1968	Systems Theory (Daniel Katz and Robert Kahn / W. Buckley)
1967 / 1973	Contingency Theory (Paul Lawrence and Jay Lorsch / Jay Galbraith)
1963 / 2001	Organizational Learning (R. Cyert and J. March / M. Dierkes *et al*)

B.4.1. Organizations like Instruments

This subsection presents three schools of organizations and management. They are symbolically classified as "Organizations like Instruments" because they concern the organization as a complex device which can operate with maximal efficiency. Little importance (if any) is given to human needs, motives, behaviour and to the cognitive limits of the participants in the organization. Moreover, these schools are alike in that they give little attention (if any) to the environment.

- *Bureaucracy School*

The bureaucracy school is one of the approaches to the study of organization structure. It was found by the social scientist Max Weber in the first decade of the 20th century. Nevertheless, Weber's works on bureaucracy just became well known after their translation from German into English in the 1940's [Pugh, 1997].

Weber's works on organizations include the description of administrative structures characterized by three different forms of authority [Scott, 1998]. Put shortly, they are traditional authority, which relies on sanctity believes and traditions; charismatic authority, which is synonymous with the ability of people to influence others; and rational-legal authority, which is established according to the normative rules of a social structure.

Although Weber's works are considered broad enough to encompass the analysis of administrative structures of different types and cultures, his theories were mainly proposed to investigate structures which could better serve and explain organizations of west civilizations. Among his analyses, Weber emphasized that in modern society of continuous growth of rationalization, and capitalism, the bureaucracy form of organization had become preponderant because of its technical superiority and greater efficiency. Therefore, rational-legal authority becomes the main structure of analysis within the bureaucracy school.

First and foremost, bureaucracy is synonymous with hierarchy of authority and it includes: delegation of authority; specification of roles and departmentalization; division of labour and specialization; payment of wages based on hierarchical positions; reward systems based on technical skills and knowledge; and normative rules and standard procedures. Following such attributes, the organization of the bureaucracy school appears to be efficient, rational and robust, with this latter representing the capability of the organization to replace its participants without major effects over its whole structure and performance.

Nevertheless, although being one of the pioneer schools with distinguished contributions to organization theory, the bureaucracy school has received many critics. Among them, the literature has mentioned that this school regards the organization as a type of instrument designed to pursue its maximal efficiency, and to achieve goals. The participants in the organizations are assumed to be passive instruments, and thus little attention (if any) is given to their motives, rational limitations and social behaviour. Moreover, little emphasis (if any) is attributed to the environment.

- ***Scientific Management School***

The scientific management though emerged during the apogee of the Industrial Revolution in the United States of America, in the late nineteenth and early twentieth centuries [George, 1972]. It developed mainly from the prominent work of Frederick Taylor on The Principles of Scientific Management [Taylor, 1911]. Taylor was an engineer from Philadelphia with pragmatic ideas on both shop floor and management activities.

The scientific management school is mainly concerned with the rationalization of technical and managerial tasks. It invokes the replacement of rules of thumb with scientific methods which are supposed to provide organizations with measurements and standardization procedures of maximal efficiency, i.e. the production of maximum output through minimum input of resources.

The scientific management though provided the field of organizations with important contributions and new requirements, among those of:

- The development of procedures for hiring and training the participants in the organization.
- The payment of higher wages in return to higher productivity.
- The specification of methods of management, like planning, coordination and control.
- The equal division of work between technical and management activities.
- The introduction of standard procedures and products; the concepts for quality control; and the basic principles for the subsequent era of mass production.

Although accepted by many industrials as an innovative approach for management, the scientific management school had its principles rejected by many workers, managers and researchers who did not accept the idea of an organization functioning as an efficient instrument of rational procedures [Scott, 1998]. Similarly to the Weber's school, the scientific management gives little attention (if any) to human needs, social behaviour and cognitive limitations; and thus it considers the participants in the organization as passive instruments. Moreover, little emphasis (if any) is given to the environment.

- ***Administrative Theory School***

The administrative theory school was developed simultaneously with scientific management. Among its proponents was the French industrialist Henry Fayol. However, his work on General and Industrial Management (published in 1916) was only translated from French into

English in the late of 1940's [Scott, 1998].

The administrative theory school is concerned with the prescription of general principles of management and it includes [Pugh, 1997]: division of work (specialization, role specification and departmentalization); authority and responsibility; discipline (obedience to general agreements between the organization and its participants); unity of command (each employee should receive orders from one superior only); unit of direction (the use of a single plan to coordinate a group of activities having the same objective); subordination of individual interest to the general interest (the interest of the organization should come always before that of its single participants, groups and units); remuneration of personnel (including reward systems, and also different methods of payment: time-rates, job-rates and piece-rates); degrees of centralization and decentralization; scalar chain (hierarchy of authority); order (the efficient allocation of human resources in the social positions of the organization); equity (a combination of justice and kindliness); stability of tenure of personnel (the minimum period of an employee to learn the activities of his new job and to give successful results in doing it); initiative (proactive); and e*spirit de corps* (union of the participants in the organization).

The Fayol's school of administrative theory along with the Taylor's principles of scientific management and the Weber's theory on bureaucracy have similarities in the sense that they concern the organization as an instrument which can operate with maximal efficiency, no matter how much significant are human needs, motives, behaviour and the cognitive limitations of the participants in the organization. Moreover, they are alike since they give little attention (if any) to the environment (e.g. its influence on the organization, its dynamics and resources). However, while the theories of Fayol and Weber analyse the organization from a top-down approach, at the structural level of analysis, the Taylor's school establishes principles to rationalize the organization from a bottom-up perspective [Scott, 1998]. Additionally, the work of Fayol on administrative theory is more concerned with the design of processes - among those of planning, specialization, leading, coordinating and controlling - which cut across the structure of the organization; while Weber's work on bureaucracy is more related to the design of the anatomy or structure of the organization.

B.4.2. Organizations like Behavioural and Cognitive Processes

This subsection presents two schools. The first is called human relations, and it concerns not only behavioural and social aspects of people in organizations, but also ergonomics. The second is called administrative behaviour and decision-making school, and it involves not only the study of behaviour in organizations, but also some cognitive processes within the organization. These schools do not resemble organizations as instruments, but they concern the organization as a system which can operate with efficiency when behavioural and cognitive processes are understood, planned and coordinated.

Although representing a breakthrough in the field of organizations, the human relations movement gives little (if any) attention to the environment. However, it was the school of administrative behaviour and decision-making that gave new insights on the importance of the environment as a stimulus to the organization response and behaviour.

- *Human Relations School*

The human relations school flourished in a period between the 1920's and 1940's, simultaneously with the development and apogee of theories on behavioural and social psychology. Although it received contributions from the analyses of various researchers, the school of human relations was greatly influenced by the work of the industrial psychologist

Elton Mayo [Scott, 1998].

This school conducted much of its work from the results of a series of studies carried out at the Hawthorne plant of the Western Electric Company, located in the western suburbs of Chicago, during the late 1920's and early 1930's. These studies were accompanied by various experiments based on stimuli-responses and reward methods in order to analyse workers' behaviour and productivity under different circumstances.

Among these studies are: the Illumination Experiment; the Relay Assembly Room; and the Bank Wiring Room [Vecchio, 1995]. The first and the second studies were intended to analyse whether changes in the work setting would influence the behaviour and productivity of the employees. The third study was intended to analyse the social behaviour of groups under the eyes of a supervisor.

The Illumination Experiment was concerned with the effects of different intensity of light (stimuli) on the behaviour and productivity of workers (responses). The researchers could observe an increase in productivity not only in the rooms of high intensity of illumination, but also in those rooms with low intensity of light.

Within the experiments of the Relay Assembly Room, the investigators introduced new conditions in the work setting of a group of female employees such as rest periods, free mid-morning lunch, a five-day workweek and variations in methods of payment. As a general result, the investigators observed a gradual increase in productivity over the course of the entire study, and also a lower rate of absenteeism.

The Bank Wiring Room was concerned with the investigation of the behaviour of a group of men in their standard conditions of work, but when supervised by an observer. The investigators could notice that informal structures of behaviour had emerged and became patterned among the workers of the group under supervision. Such behavioural structures were developed (and shaped by internal conflicts) in order to regulate the behaviour, the attitudes and thus the productivity of the workers. Hence, no disparity of productivity could be observed by the supervisor.

Conclusions on these studies showed that:

- Organizations are arenas of emotions and feelings.
- Participants in organizations have motivations which are influenced not only by economic interests, but also by their values, sentiments and by the behaviour of others.
- Organizations shape participants' behaviour (from a top-down view), but participants also shape organizations (from a bottom-up view).
- Normative structures shape behaviour as behavioural structures shape norms.
- Leadership characterized by attention and equality can give better results than authoritarian structures.

The school of human relations was criticized in different ways. One of the main reasons lies in the lack of evidence about the relation between worker satisfaction and productivity. However, in its broad sense the human relations school represented a re-orientation on the research of management and organizations, since it provided an initial background towards the discipline of organizational behaviour.

- *Administrative Behaviour and Decision-Making School*

The administrative behaviour (and decision-making) school emerged during the 1940's with the prominent work of Herbert Simon who was awarded in 1978 the Nobel Prize in Economics [Simon, 1997b].

In short, administrative behaviour describes and explains organizations in terms of cognitive processes (with special attention for decision-making). This school is less prescriptive than the previous schools. It describes management and organization processes as they really like to happen in practice, and not what and how they ought to be.

The central contribution provided within the school of administrative behaviour is concerned with the concept of bounded rationality and its implication for organization behaviour. In short, bounded rationality designates theories of rational choice which recognize the cognitive limitations of the decision maker - limitations of both knowledge and computation. Therefore, the concept of satisfying (satisficing) criteria and goals is applied instead of optimizing (or maximizing) them. In such a way, theories of bounded rationality extend classical economic theories of rational choice to a more realistic perspective on human decision-making as supported by cognitive psychology research. Among the implications of bounded rationality for the literature of social sciences are the development of the discipline of behavioural economics [Simon, 1997a] and other branches combining psychology and economics [Rabin, 2002].

Simon's work on administrative behaviour was influenced by some other authors on management science [Barnard, 1938]; decision-making in organizations [March, 1994; and March and Simon, 1958 and 1993]; artificial intelligence and cognitive science [Newell and Simon, 1972]; and it has also influenced many other researchers of different areas raging from artificial intelligence, psychology and computer science [Carley and Gasser, 1999] to economics [Cyert and March, 1963].

In 1958 March and Simon authored a remarkable book titled "Organizations", which was latter revised in 1993 [March and Simon, 1958 and 1993]. They introduced concepts of administrative behaviour into organizations, and then they proposed a new theory on organizations. Some of their contributions are enumerated in the next paragraphs.

Firstly, March and Simon viewed organizations like vertical and horizontal structures constituted by sub-goals and goals. Sub-goals are formed at lower levels and they represent means for the achievement of more complex goals (ends) at upper levels.

Note: In such a view of structures of goals and sub-goals, the organization selects and allocates specific sub-goals for its participants and units; and such a procedure can be understood as synonymous with the cognitive process of attention which plays the role of directing and focusing certain mental efforts of the participants in the organization to enhance perception, performance and mental experience during task execution. Moreover, high mental processes of cognition (such as decision-making and problem-solving) play an important task in the organization by providing means for planning and achievement of sub-goals, moving towards more complex goals. Hence, the processes of attention (selection) and division of work (specialization) in the organization can reduce the amount of information which has to be processed by the participants. Consequently, these processes reduce the amount of uncertainty within the organization which emerges from the more complex goals at upper levels and from the influence of the environment.

Secondly, March and Simon provide the literature with the perspective of the organization as a set of programs which can be evoked due to an individual, organizational or environmental stimulus. These programs are classified into programmed and non-programmed decisions [Simon, 1977]. The former type plays an important part in the processes of coordination of recurring or repetitive tasks; and the latter applies to planning, innovation and learning practices in organizations. Since programmed decisions are recurrent programs for repetitive stimulus, they provide the organization with predictability; they reduce the amount of information to be processed by individuals when taking decisions and solving problems; and thus they contribute to reduce the amount of uncertainty that the organization confronts.

Thirdly, March and Simon has also provided the literature with the perspective of the organization as an information-processing system constituted by distributed computational minds and agents. In such a view, communication plays an important part by: transmitting decisions from one agent to another; channelling information; and controlling the flow of decisions. This perspective has been further explored and extended to a new research on computational organization theory [Carley and Gasser, 1999; and Prietula *et al*, 1999].

The school of administrative behaviour not only represents a re-orientation in organization theory, but also the emergence of a body of concepts which play the most important part in the functioning of organizations of today. Among these concepts is the perspective of organizations as cognitive systems with the ability to learn.

Re-orientation concerns mainly the transition of emphasis given to the participants in the organization. The schools of bureaucracy, scientific management, administrative theory and human relations do give attention to either the structure of organizations or the management processes within the organization, but they seriously understate the cognitive abilities and limitations of the participants in the organization. The administrative behaviour school emerged to extend these schools to a theory which describes and explains organizations (and thus their structure, management processes and behaviour) in terms of processes of cognition - e.g. perception, attention, memory and communication, decision-making, problem-solving and learning.

—

Note: Processes of information or knowledge management play the most important role in organizations of today. These processes comprise cognitive tasks of [Reed, 1988]: selecting information (perception and attention), recording (memory), forming and organizing knowledge (concept formation and categorization), information-processing (decision-making), planning and innovating (problem solving and learning) and acting (response). Hence, theories of administrative behaviour play an important part in organizations of today; no matter whether among the participants in the organization are machines that pursue cognitive abilities for autonomous and intelligent action.

The school of administrative behaviour gives some attention to the environment and it treats it mainly as a source of stimulus for evoking programs, decisions and actions within the organization. However, this school gives more emphasis on the individual and the organizational levels of analysis. The subsequent schools of systems theory, contingency theory and organizational learning emerged to give greater attention to the environment, and then to open new perspectives on organizations, the environment and population of organizations.

The school of administrative behaviour flourished in a period of both: transformation of psychology research and apogee of general systems theory. It is highly probable that Herbert Simon was influenced by the new theories of cognitive psychology which emerged during the

1950's; and most importantly by the advent of digital computers which played an important part as metaphors for the development of a theory on information-processing systems [Newell and Simon, 1972; Reed, 1988; and Reisberg, 1997].

In another context, bounded rationality and general systems theorists played a counterpart task in the literature by proclaiming the lack of mathematical tools for coping with more complex problems (where human behaviour, emotions and cognition are key factors). On the one hand, the theory of bounded rationality [Simon, 1997a] called for new approaches which could extend the methods of statistical decisions analysis used by economic theorists to a more realistic scenario about human decision-making [Simon, 1997a]. On the other hand, general systems theory pointed out the need for new mathematics in order to narrow the gap of understanding between the analysis of non-living and living systems [Zadeh, 1962]. This research advocates that such a new approach (as proclaimed by bounded rationality and general systems theorists) emerged with the advent of fuzzy systems theory [Zadeh, 1965 and 1973] and its derivatives on computing with words and perceptions [Zadeh, 1996a and 2001].

Administrative behaviour (within decision-making) plays an important role within the concepts proposed in Part II.

B.4.3. Organizations like Systems: Analysis and Design

The schools presented in this subsection give special attention to the environment. They are called schools of systems theory, contingency theory and organizational learning. They emerged during the second half of the 20th century and they were mostly inspired by many of the concepts proposed and exploited by general systems theorists. The core of these schools is the analysis and design of organizations.

- *Systems Theory School*

The school of systems theory provides a picture of organizations as systems[48] with input and output relations to the environment[49] [Silverman, 1970].

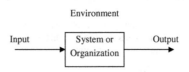

Figure B.1. Input and Output Relations between the Organization and the Environment

-----

[48] The subjects of systems and environment are further described in Part II. Nevertheless, one provides a short definition of them herewith as proposed in [Hall and Fagen, 1956]: "A system is a set of objects together with relationships between the objects and between their attributes" (or properties).

[49] "For a given system, the environment is the set of all objects a change in whose attributes affect the system and also those objects whose attributes are changed by the behaviour of the system".

Systems theorists have the perspective on the environment as a source of resources which are necessary for the organization survival, evolution and development. The environment shapes the organization and it influences the elements of the organization (such as its social structure, technology and goals along with the motives, perception, emotions and behaviour of its participants). Moreover, organizations also influence the environment; they influence the economic, political and social contexts of the environment.

Systems theorists have attempted to describe organizations as processes of organizing which fall in two main domains. The first domain concerns those processes which lead the organization to adapt to its environment; and they involve the concepts of learning, information-processing, open systems, entropy, self-regulation and homeostasis [Scott, 1998]. The second domain involves the dynamic processes of interaction between the elements of the organization and they can be stimulated by internal and external sources. Hence, a change in one or more of the elements affects the whole organization. Systems theorists also regard organizations as hierarchical and loosely coupled systems [Scott, 1998].

Not only have the principles and concepts within general systems theory influenced the school of systems theory, but also the mathematical and computational tools which emerged during its development and apogee. Among the mathematical background used to the analysis of general systems (and thus organizations) are theories of linear and non-linear systems, stochastic and learning systems, optimal systems and fuzzy systems [Zadeh and Polak, 1969; and Zadeh, 1973]. Some of these mathematical concepts (supported by computer simulation) have been largely used for instance by The MIT System Dynamics Group of the Sloan School of Management since the 1960's. The main purpose of this group has been the analysis and design of complex systems (including organizations), and also the prediction of behaviour of industrial, social and general systems [Forrester, 1961 and 1973].

With the advent and popularization of digital computers, new tools emerged in order to support the analysis and design of processes and systems of higher order of complexity. Among these tools are distributed artificial intelligence [Bond and Gasser, 1988], software agents and multi-agent systems [Nwana and Azarmi, 1997; and Weiss 1999] and soft computing [Zadeh, 1994]. Such mathematical and computational tools have stimulated the development of new approaches for the: simulation of organizations and social networks; analysis and design of organizations; and test-generation of propositions and theories of organizations. Following this direction, the multi-disciplinary centre for Computational Analysis of Social and Organizational Systems of the Carnegie Mellon University emerged to carry out research towards the discipline of Computational Organization Theory (COT) [Carley and Gasser, 1999; and Prietula et al, 1998].

Among the prominent contributors to the school of systems theory are the names of [Khandwalla, 1977]:

- D. Katz and R. Kahn, who provided a theory about the cycle of input-throughput-output in organizations.
- T. Parsons, who has defined the technical, managerial and institutional subsystems of the organization.
- F. Emery and E. Trist, who defined the organization as a system composed by subsystems (divisions) which are influenced by technical, economic and social forces.
- H. Leavitt, who described a model of the organization and the process of interaction between the elements of the organization. A change in one or more of these elements

147

affects the others and thus the whole organization.

Although providing organizations with new insights on the environment and with general principles of organizing and models of organization, the school of systems theory did not escape from criticisms. Most of the critics advocate that the concepts, principles and theories of organizations provided by the school of systems theory are too generalist to address the particular features of different organizations which are contingent on the situation or context in which they exist.

Systems theory (within the broad context of general systems theory) plays an important role in Part II.

- *Contingency Theory School*

The core of contingency theory is concerned with the design of organizations and its basic premises assume that [Galbraith, 1973 and 1977]:

- "There is no one best way to organize".
- "Any way of organizing is not equally effective".

The first premise asserts that there is no general principle for organizing. Instead, principles of organizations and processes of organizing are contingent upon the situation, context or environment. The second premise complements the first one and it asserts that different structures of the organization and different processes of organizing lead the organization to different results (subsuming behaviour, performance, efficiency, efficacy, and so on).

The practice of designing organizations can be traced back to ancient and medieval civilizations, but it only developed as a discipline of organization theory during the 20th century. The schools of organizations previously presented in this section were the first schools to provide principles of design to organizations. These principles can be classified in structural and processes design. However, although most of the previous schools had developed principles of organization and organizing which were contingent on particular contexts, they published their results in the form of general concepts, propositions and theories. These generalizations mark the foremost line of separation between previous schools of organizations and contingency theory.

The school of contingency theory concerns the organization (and thus its elements), the features of the organization (like its size) and also the processes (of management) as objects of design which are dependent upon the environment (i.e. its cultural, technological, economic, political and social contexts). Moreover, contingency theory concerns the organization as a dynamic system (which changes over time). A change in one of the elements of the organization influences the other elements. Hence, the organization needs to be redesigned periodically.

Contingency theory is not only a qualitative and prescriptive school, but also quantitative and descriptive. Many of its results are derived from large-scale empirical research. Although being classified as distinct, the approach to comparative studies (or comparative analysis) of organizations has provided and influenced the school of contingency with a broad spectrum of practical results [Blau, 1981; Blau and Scott, 1963; and Scott, 1998].

Among the prominent contributors to the school of contingency theory are the names of [Galbraith, 1973; and Khandwalla, 1977]:

- T. Burns and G. Stalker, who observed two distinct forms of management in a set of 20 British and Scottish firms. These were called mechanistic and organic styles. Burns and Stalker identified that both styles were effective, but in different contexts. The mechanistic form was associated with a more stable environment and the organic form with a more dynamic environment of technology innovation.

- J. Woodward, who studied 100 British firms, observed relationships between the form of technology used by the organization and its structure. Among the results, Woodward concluded that different firms with different levels of technological complexity of production had distinct spans of control[50] and distinct hierarchies of authority, with the latter being characterized by the number of managerial levels in an organization of scalar principle[51].

- J. Thompson, who, among other contributions, proposed that the different levels or layers of the organization can have distinct behaviour, and thus distinct social structures. Thompson uses the definitions of organizational levels of analysis as proposed by Parsons (1960) and he explains that [Scott, 1998]: - on the one extreme is the technical level which resembles the concept of rational systems. This level is more like a closed system of normative structure which tries to protect the organization from uncertainties of the environment; - on the other extreme is the institutional level which resembles the concept of open systems. This level relates the organization to its broad environment; in the middle is the managerial level which resembles the concept of natural systems. This level is more like a system of behavioural structure with enough flexibility to mediate tasks between the two extreme levels.

- P. Lawrence and J. Lorsch, who coined Contingency Theory [Scott, 1998], proposed that organizations face distinct degrees of differentiation and integration of activities. Differentiation of social structure and technology between and within the various divisions of the organization is contingent upon the distinct tasks environments that they face. Task environments vary along a continuous scale of complexity, subsuming market dynamics (innovation and variability of goods and services), uncertainty and predictability. For example, in some industries, research and development departments can face a more complex environment than do production units. However, differentiation requires integration. The greater the differentiation among the subtasks of the organization, the more complex is the integration in order to achieve successful completion of the whole task. Lawrence and Lorsch carried out empirical research with ten organizations in three industries. Not surprisingly, they confirmed their propositions and predictions about differentiation and integration.

Some other authors could join the list of contingency theorists, such as A. Chandler and

---

[50] Span of control refers to the number of employees or subordinate roles which can be effectively supervised and coordinated by a first line manager or superior [Galbraith, 1977; and Scott, 1998].

[51] The scalar principle states that the participants in an hierarchical organization are linked by single relations and the flow of decisions and authority should move from upper layers (of management) downwards lower layers (of workers) [Galbraith, 1977; and Scott, 1998].

R. Hall [Galbraith, 1973], and C. Perrow [Khandwalla, 1977]. However, the conclusions would be almost the same: the best way to organize is contingent upon the uncertainty and diversity of the tasks performed by the organization and by its divisions [Galbraith, 1973].

Contingency theory plays an important part within the concepts introduced in Part II. Most of the contents of Part II about organization design are based on the work of Galbraith [1973, 1977 and 2002].

- *Organizational Learning School*

Organizational learning is a multi-disciplinary field of research which focuses on creation and management of knowledge [Dierkes *et al*, 2001]. It has received contributions from diverse areas ranging from psychology, management science, sociology and anthropology to philosophy, history, political science and economics.

As recognized in the literature [Dierkes *et al*, 2001], the pioneers in the field of organizational learning include Argyris and Schön [1978], Cyert and March [1963] and March and Olsen [1975]. However, this work puts forward the perspective that organizational learning was firstly touched upon in the work of Simon [1947] and March and Simon [1958] on organizations, administrative behaviour, decision-making and bounded rationality. The point of departure for such an assertion is that Simon and March [1958] also regarded the organization as a system with cognitive processes and with the ability to learn new routines and programs towards innovation.

Agents of organizational learning involve the participants within the organization and the relationships or social networks which they form. Learning in organizations is also supported by the goals, technology and social structure of the organization. Moreover, organizational learning is also influenced by inter-organizational processes and thus by the environment.

In psychology research, learning is defined as the process of making changes in the working of our mind, behaviour and understanding through experience [Bernstein *et al*, 1997; and Minsky, 1986]. This work borrows such a definition to assert that:

Definition B.4.3.1: Organizational learning involves a process of making changes in the behaviour of the organization through experience.

A process of organizational learning involves an adaptive learning cycle which is in fact a concept borrowed from principles of feedback control and adaptive systems of the broad field of cybernetics and general systems theory [Buckley, 1968; and Wiener, 1948, 1954 and 1961]. This cycle operates as either single-loop learning or double-loop learning. On the one hand, with single-loop learning the organization identifies problems and makes corrections by using previous solutions and present procedures. On the other hand, with double-loop learning the organization identifies problems and makes corrections by changing present procedures, policies and processes [Daft and Noe, 2001].

Organizational learning is also synonymous with continuous process improvement [Paulk *et al*, 1994] and in such a way it plays an important part in the understanding of the industrial case results presented in Part III.

B.5. Perspectives on Organizations

The schools of organization theory, as developed during the 20th century, have proposed different definitions of organizations which lie in one of the perspectives among rational, natural and open systems [Scott, 1998]. Additionally, some of these definitions can be better classified as a combination of such perspectives.

B.5.1. Rational Systems

Rational systems theorists assume that:

- Organizations are social systems of interacting agents.
- Organizations pursue specific goals.
- Organizations are based on high degrees of formalization.

These premises together classify organizations as special types of social systems which are characterized by goal specificity and formalization. Both characteristics are purposefully achieved by the processes of design.

Organizational goals are specified within the process of strategic planning and they provide the organization with direction and criteria of choice (among products and services, market of action, structure, processes and technology, participants, etc.).

Formalization is achieved with the specification and implementation of a normative structure which regulates the behaviour and decisions of the participants in the organization. Formalization requires the activities of specialization (distribution of roles), span of control (number of agents coordinated within a division or superior), hierarchy of power, centralization and departmentalization.

Rational systems theories comprise the concepts of design and redesign. The former can be understood as a mean for the achievement of efficient organizations, and the latter, for the improvement of organizational performance (including social, cognitive, institutional and economic factors).

B.5.2. Natural Systems

Although natural systems theories were developed to contrast the emphasis on rationality advocated by rational systems theorists, the former can be understood as a complementary perspective to the latter.

Natural systems emerge when the rational concepts of goal specificity and formalization are relaxed and extended to a more complex definition which asserts that:

- Organizations are social systems of interacting agents.
- Organizations pursue multiple goals.
- Organizations evolve informal and behavioural structures.

Like rational systems, natural systems theorists do consider organizations as a class of social systems, but of a different kind from that of rational systems. Emphases are played on goal complexity and informal structure, rather than on goal specificity and formalization.

Goal complexity characterizes the multiple goals and interests of the agents (or participants) of the organization. Although organizations may have specific goals, the behaviour of their participants may be governed by their own motives. Among them there exist common and divergent interests, and thus cooperation and conflict phenomena.

Informal structure emerges from social relationships and coalitions among the participants in the organization; and it evolves to a patterned behavioural structure which differs from the rules prescribed within the normative structure.

Natural systems theorists also advocate that rationality constrains the creativity of the participants in the organization, since it regulates behaviour and decisions. Most importantly, they view the organization as an indispensable resource which evolves and adapts to particular situations according to internal forces; moreover, the organization attempts to survive and to support the interests of its participants. However, if survival is under threat, then new patterns of behaviour emerge among the participants and their individual and collective interests suppress the specified organizational goals.

B.5.3. Open Systems

The open systems perspective plays emphasis on both: the dynamics of the organization and the relation between the organization and the environment. It assumes that:

- Organizations are social systems of interacting agents.
- Organizations are dynamic systems coupled with the environment.
- The environment shapes organizations (and vice-versa).

Open systems theories extend the perspectives of rational and natural systems to a broader concept of organizations as systems connected with major systems (or supra-systems). The definition of the organization becomes more complex because: the border between it and the environment is more a less vague and fuzzy; the elements of the organization are dependent upon each other, and a change in one of them influences the whole organization; agents join the organization and participate in the activities of the organization if they can reach satisfactory bargains; and as the organization is open, it is subject to the uncertainty of the environment. However, the environment is viewed as a source of resources for the organization which is necessary for its evolution, development and survival.

B.5.4. Open-Rational Systems

The open systems perspective motivated some authors to review their original theories on (closed) rational systems, and influenced other researchers to develop theories on (opened) rational systems. Such a fusion of perspectives gave origin to the concept of open-rational systems. Among the theories which can be classified within open-rational systems are administrative theory (or bounded rationality), comparative structural analysis, and contingency theory [Scott, 1998].

B.5.5. Open-Natural Systems

Similarly, open-natural systems represent a fusion of the perspectives of natural and open systems. Among the concepts within this perspective are those of: evolutionary processes, adaptation, survival, natural selection, institutional and social stability, and belief networks [Scott, 1998].

B.6. Rationales for Organizing: Political, Economic and Social Contexts

Organizations are important for many reasons, and perhaps the most ancient rationale for organizing was the one stated by Aristotle (384-322 B.C.), i.e. "the whole is more than the sum of its parts" [Wren, 1987].

Broadly speaking, organizations provide people with processes. These processes support them in the achievement of complex goals. Such complex goals have political, economic and social facets.

B.6.1. Political Facet

Organizations provide people with power - in the form of legitimate or informal authority [Scott, 2001]. Power resides in people's dependencies [Kipnis, 1990], and organizations gain in power as people depend on them, says, information, goods, services, wages, rewards and so on. It can be created in diversified forms: by agglomerating muscles as explored during the prehistoric era, ancient civilisations, Middle Ages and later in lesser scale after the Renaissance; and also by assembling individual and collective cognitive processes in order to achieve complex information-processing systems to cope with the uncertainty of the environment. The latter has been increasingly explored after the apogee of the Industrial Revolution in Europe and in the United States of America. Nevertheless, it started to receive more attention and new insights from the last fifty years only, empowered by the advances in computers, communication networks and artificial intelligence research.

B.6.2. Economic Facet

Organizations provide people with efficient costs of production and transactions - the transfer of goods and services from one individual to another [Milgrom and Roberts, 1992]. This facet supports the latter by providing organizations with capabilities to create incentives to people, to the market and to other organizations.

B.6.3. Social Facet

Organizations provide people with sociability, societal status, knowledge, satisfaction, physical and mental health, and general well-being. The economic facet supports this one since the history has shown that the emergence of the new organizations after the Industrial Revolution has provided people with better standard of life and with longer expectation of life [Johnson, 2000; Richard, 2000; and Wren, 1989]. Nevertheless, such positive effects were also supported by other transformations like in the transport, urbanisation, sanitation and medical affairs. The social part plays important roles in both political and economic facets. Through charisma and status for instance, people may relate to, and infiltrate into political and economic decisions.

People join organizations for different reasons and motives, which may have political, economic and social facets. The facets briefly presented in this section form a framework to support a rationale for people to participate in organizations.

B.7. Benefits of Organizations

Organizations can benefit people politically, economically and socially, as discussed in the previous section. Moreover, organizations benefit people if they are considered as means to overcome some limitations of their individual agents. Such limitations are classified as cognitive, physical, temporal and institutional [Carley and Gasser, 1999], and they can also be extended to include *spatial limitation*, as introduced here.

B.7.1. Cognitive Limitation

Firstly, agents have cognitive limitations according to the definition of bounded rationality [Simon, 1997a and 1997b; and March and Simon, 1958]. Organizations can improve such limitations by providing them: with the support of additional cognitive processes of other cooperative agents; with a social structure to support decision-making and problem solving; with common goals and strategies in order to simplify choices and alternatives; with division of work and problems, through departmentalization and decentralization for instance; with the formation of channels of communication to spread and share knowledge; and with technology which provides automation of cognitive tasks - like those of communication, computation, storing and memory, decision-making and problem solving - and other technologies - like those of process improvement and quality programs - to name some but a few of them.

B.7.2. Physical Limitation

Secondly, agents have physical limitations due to their physiology. Agents cannot carry weight as much as they may need; they cannot move as fast as they may want; and they cannot handle as many things as they may desire. Organizations can provide them with technology and coordination in order to support their physical capabilities.

B.7.3. Temporal Limitation

Thirdly, agents have temporal limitations and therefore they have to organize themselves together to achieve goals which transcend the lifetime of any one agent.

B.7.4. Institutional Limitation

Fourthly, Agents are legally or politically limited and therefore they have to attain organizational status in order to act as a corporate actor rather than as an individual one.

B.7.5. Spatial Limitation

Fifthly, agents are spatially limited because they cannot be in different locations at the same time. Therefore, organizations can provide them with distributed resources, and also distributed capabilities of cognition and action in the form of a multi-agent system.

Organizations also provide intelligence to the connections or relationships between their agents. Such intelligence cannot be found with one agent only, and neither by simply putting together some agents.

Organizations as a whole pursue robustness. They can still keep the same behaviour and performance even when one or more of their agents are replaced with others.

B.8. Elements of Organizations

Organizations have received diverse and complex definitions in the literature. Hence, it may be helpful to establish a simplified model focusing on their major elements. The organizational model illustrated in Figure B.2 was firstly proposed by Leavitt [1965]. Nevertheless, it was later stretched by Scott to include the environment as a separate factor which is regarded as an indispensable ingredient in the analysis of organizations [Scott, 1998].

This research supports the definitions of the elements of this model as presented by Scott, and it also extends them in order to provide foundations for the scope of organizations and technology introduced in Chapter 1. Major extensions are proposed to the concepts of the participants in the organization since they include not only people, but also *cognitive machines*. The influence of new participants on the definitions of the other elements of the organization is also taken in consideration. Nevertheless, a better description of the subject of *cognitive machines* in organizations is presented in Part II.

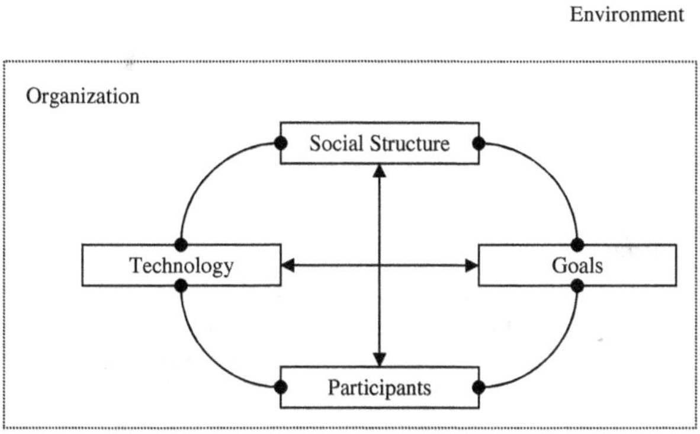

Environment

Figure B.2. A Model of the Organization

B.8.1. Participants

The participants in the organization are agents and they include people, other organizations and machines. People and other organizations are those individuals and social groups who, in return for a variety of inducements, make contributions to the organization [Scott, 1998]. People can be synonymous, for instance, with employees, managers, customers and stakeholders. By other organizations it is meant subcontractors, buyers, suppliers and other partners of the central organization.

Machines are agents which act in the name of organizations, as people do. Therefore, they fulfil roles in the organization ranging from muscular and repetitive activities to complex cognitive tasks. Hence, they can fulfil roles in organizations at the technical, managerial, institutional and worldwide level.

B.8.2. Social Structure

Social structure refers to the standards and regularized aspects of the relationships existing among the participants in the organization; it comprises two types of structure, called normative and behavioural structure [Scott, 1998].

* *Normative Structure*

A normative structure is an institutionalized set of rules which state what ought to be the

155

behaviour of participants in the organization. Therefore, it provides reductionism and puts constraints on the behaviour of the organization. However, it can reduce the amount of uncertainty in the organization not only by channelling and equalizing information among its participants [March and Simon, 1993], but also by governing and patterning behaviour and decisions.

A normative structure includes norms, values and roles. Put shortly, values are criteria used in selecting goals of behaviour; norms are rules governing behaviour which specify means, strategies and tactics for pursuing goals; and roles are expectations of behaviour for the occupants of specific social positions[52]. Roles also represent similarities in the responses of different individuals to a common situation [Khandwalla, 1977]. Therefore, in normative structures, values, norms and roles are organized to form a consistent set of prescriptive rules to govern the behaviour of participants.

Moreover, normative structures also play an important part in governing the processes of decision-making in organizations. Such decisions are synonymous with programmed decisions, since they are derived from detailed prescriptions which govern the sequence of responses of the organization (or divisions of the organization) to the environment [Simon, 1977].

- *Behavioural Structure*

Behavioural structure focuses on actual behaviour rather than on prescriptions for behaviour [Scott, 1998]. Analysis of behavioural structure concerns those activities, relationships and emotions which exhibit some degree of regularity and similarity over a period of time. These can be among individuals, social groups and networks of organizations.

Behavioural structures also play an important role in processes of decision-making which belong to the class of non-programmed decisions [Simon, 1977]. They are non-programmed because they are not governed by a set of rules and thus they have no specific prescriptions to deal with a situation like the one at hand. Instead, participants must use and explore their general capabilities of cognition, intelligence and autonomy for decision-making and problem-solving. Given a large spectrum of situations and problems, no matter how ill-structured and novel they are, man has a remarkable ability to solve them. The decision of a company to establish operations in a continent where it has not been before is an example of non-programmed decision[53].

Normative and behavioural structures of organizations are neither completely aligned nor totally independent of each other. Instead, they are better represented by an intermediate degree of compatibility and similarity of behaviour. Moreover, behaviour shapes norms just as norms shape behaviour [Scott, 1998].

B.8.3. Goals

Goals have received controversial opinions regarding their importance, and also different definitions in the literature of organizations [Scott, 1998]. Nevertheless, goals constitute an

---

[52] A social position is simply a location in a system of relationships [Scott, 1998].

[53] Nevertheless, non-programmed decisions also happen daily in people's life. Familiar examples are parking a car, driving in city traffic, cooking a different meal, summarizing new stories, and so on.

important element with this research. Therefore, a definition is introduced in the following which is further discussed in this research.

Firstly, goals are ends, but they also can represent means to the achievement of other goals. Nevertheless, strategy and tactics are better definitions of means to the achievement of goals [Galbraith, 2002].

Secondly, goals can be applied to different levels of analysis of the organization. From a top-down path, these levels are environmental, institutional, managerial and technical systems.

Proposition B.8.3: The better the alignment of the goals of the organization, from a top-down perspective, the greater is the probability to satisfy them.

Third, participants have goals which differ from those of the organization. Some authors call them motives [March and Simon, 1993], and they represent the desires of participants to join and to act in the name of the organization.

Fourth, if participants have motives and they include machines, then: - do machines have goals (motives)? The answer is yes if they are considered to carry people's desires, motives and general goals of the organization. It applies mainly for those machines which fulfil cognitive roles in organizations. A simple example is the elevator. It has the aim of transporting people from one floor to another according to a procedure of programmed decisions embedded in its digital memory. Another example, and more complex, requiring a higher level of cognition, is the application of electronic agents in the internet. They can fulfil the roles of searchers of information, buyers and suppliers of general services, and thus they can act in the name of the organization.

B.8.4. Technology

Technology is important and indispensable for organizations as the other elements. Organizations use technology in different ways and in distinct levels of application. They produce technology for their own consumption; they pass it to the environment; and they also consume it from the environment with the aim of survival, evolution and development. Appendix D surveys the subject of technology in the context of organizations and it also proposes definitions about this topic.

B.8.5. Environment

The environment of the organization comprises different levels of analysis and it involves technical and institutional aspects [Scott, 1998]. The technical aspect is synonymous with production systems and the institutional aspect is synonymous with symbolic and cultural factors influencing the organization. The environment of the organization can also include ecological systems and some organizations exist to work for the preservation of nature and natural resources.

The literature of organizations have given special attention to an environment which is synonymous with networks of organizations, local and global economy, technology, cultural values, normative processes and regulative systems, institutions, etc. However, with some few exceptions [Simon, 1977], researchers of organizations have attributed little or no emphasis to the natural resources of the environment.

This research supports the literature about this subject and it asserts that the environment of the organization exists, evolves and develops according to economic, social and political contexts. The concept of environment is further defined in Part II.

B.9. The Discipline of Organization Theory

Organization theory concerns the analysis and design of organizations. Analysis is the activity used to examine something in detail. It provides knowledge on the elements (or parts) of the organization; on the relations between the elements; and also on the whole organization. Organization design involves a continuous process of decision used to provide coherence among the elements of the organization. Such a process involves choices of: goals and strategy; social structure; technology and processes; reward systems; and policies of human resources.

In such a way, analysis and design are complementary tasks where the former plays an important role by providing designers with approaches to create or to change organizations.

Galbraith proposed a framework to support people in the activities of selecting alternative organization forms and assessing the likely consequences of their choice [Galbraith, 1977 and 2002]. This research views design according to his framework.

B.10. Nature and Diversity of Organizations

Organizations have been widely defined in the literature with divergent, complementary and common perspectives. Ironically, Perrow has compared organizations to a zoo with a bewildering variety of specimens [Perrow, 1974]. The differences in perspectives can vary according to the following three factors of diversity [Scott, 1998].

The first is the diversity of organizations, which is characterized for instance by their type, size and shape, the culture into which they are immersed, and their purpose. The church, firms, profit and non-profit institutions, public and private industries, universities, trade unions and hospitals are organizations whose features differ from each other. Nevertheless, divergences can also be found with the analysis of organizations of the same class. Firms of different sizes can have different social structures for instance, and thus they can provide different perspectives. A study about particular aspects of different classes of organizations is found in [McKinlay, 1975; and March, 1965].

The second is concerned with the interests and the background of the researchers, which can vary for instance from political scientists, economists, sociologists, psychologists and anthropologists to engineers and computer scientists. A tutorial about different writers of organizations is presented in [Pugh, 1997; and Pugh and Hickson, 1997].

The third is related to levels of analysis which can vary from a micro to a macro view. They are the social psychological, structural and ecological levels of analysis. The social psychological level focuses on individual participants and their interrelations, and it tries to explore the impact of environment, technology and social structure on the behaviour of individuals. At the structural level, the concern is to investigate the characteristics of organizational forms and design, to explain structural features, processes and the technology of organizations. At the ecological level, organizations and networks of organizations are analyzed as collective actors operating in larger and interdependent systems of relations, constituted by other organizations and their environment. These three levels of analysis are presented in [Scott, 1998] and covered in [March, 1965; and Khandwalla, 1977].

158

B.12. Formal Organizations

- *Degree of Formality*

Formal organizations have received distinct definitions spanning perspectives of economics [Milgrom and Roberts, 1992] and sociology [Blau and Scott, 1963]. Milgrom and Roberts define formal organizations as entities of independent legal identity which can sign contracts in their own name and seek court enforcement of those contracts. Blau and Scott differentiate formal organizations from social organizations stating that the first constitutes entities deliberately established for the explicit purpose of achieving certain goals, while the second emerges whenever men are living together. Despite distinct, such perspectives complement each other. The former provides the second definition with the ability of the organization to exercise contractual rights and legitimate power, while the second provides the former with the conception of organization design.

This research supports both definitions and it extends them to the principle that the frontier between informal (or social) and formal organizations is a matter of degree which depends on the choice of criteria of analysis on formalization. For example, if performance and achievement of goals are the criteria defined to measure formality among a body of organizations, then it is not possible to assure that organizations holding contracts are more formal than those which have none. Nevertheless, it may be asserted that organizations which are deliberately designed according to some criteria have higher probability of satisfy such criteria, and hence they are more formal than other organizations.

B.13. Summary

Organizations are coordinated according to social structures of normative and behavioural parts and they comprise goals (subsuming survival and development), participants and technology. Furthermore, they depend on the environment and vice-versa.

Organizations resemble cognitive systems and distributed minds. Minds are synonymous with individual agents, teams, departments or divisions; and systems are synonymous with levels of analysis (e.g. technical, managerial, institutional and worldwide systems). Organizations process, manage and exchange information between their distributed parts and also between them and the environment.

Organizations and the practices of organizing emerged during ancient civilizations and the Middle Ages, and their development was paved by a revolution in thinking, supported by religious, political, economic and social transformations during the Renaissance and the Enlightenment Ages in Europe. Such transformations created the necessary conditions for the Industrial Revolution in Europe during the 18th century, and later in the United States of America during the 19th century. The gradual maturation of organizations was encompassed by: transformations of the perceptions, behaviour and motives of their participants; evolution and development of technology; the need for new organizational processes and structures with normative and behavioural parts; the human desire of pursuing more complex goals; developments in the social sciences, and most importantly in cognitive psychology research and general systems theory; and changes in the environment.

Modern organizations emerged after the Industrial Revolution and they were challenged by new political, economic and social contexts. Thus, schools of organizations and management were developed in order to support: the analysis of the new organization and the design of new organizational structures and processes. Such schools emerged in the first decade of the 20th century, giving rise (and maturation) to the discipline of organization theory. They started with theories of bureaucracy and principles of scientific management and administrative theory, and they received new insights from the experiments of the human relations school. However, organizations of today have been closer to the contributions provided by the schools of administrative behaviour (and decision-making), systems theory and contingency theory.

Tracing back to the period of Industrial Revolution, organizations have developed from the concept of structures and agglomeration of people (with machines), to the principles of today which can be characterized by distributed cognitive processes for organizing and acting. Organization schools of today have given more attention to the concept of members of organizations (participants) as decision-makers and problem-solvers. Additionally, they understand that these participants have cognitive limitations and motives which differ from organizational goals. Thus, organizations have to provide them with inducements. Cognitive processes in organizations were mainly explored by the administrative behaviour and decision-making school. Subsequently, more attention has been given to the environment, as recognised mainly with the schools of systems theory, contingency theory and organizational learning.

The Industrial Revolution introduced new members to modern organizations, in the form of machines. Such machines challenged humans by replacing their muscular activities with mechanical and electrical mechanisms, and also by providing people with economy of time and energy (i.e. physical and cognitive efforts). People then started to give more attention to the development of additional cognitive processes for organizing.

Organizations of today have been challenged by the advent of new machines in the form of computers, software programs and communication networks. These machines (when proper designed with the background of artificial intelligence, cognitive science and systems theory) can pursue capabilities to carry out cognitive tasks in organizations. As advocated in Chapter 1, such machines are emerging to act in the name of organizations, just like people do. Hence, they enter into the arena of organizations as additional elements of design which can replace human beings when performing technical and managerial activities. However, the use of such machines requests economic, social and political analysis.

Organizations have political, economic and social facets, and thus they can provide people with power, wealth and status. Organizations are also means to overcome and to extend the limits of individual agents; such as cognitive, physical, temporal, institutional and spatial limitations. Hence, organizations benefit people in various ways.

Organization theory concerns the analysis and design of organizations. Analysis is the activity used to break something down in a minute examination of its parts. It provides knowledge of the elements (or parts) of the organization; on the relations between the elements; and also on the whole organization. Organization design involves a continuous process of decision in order to provide coherence among the elements of the organization. Such a process involves the choice of goals, social structure, technology and the participants in the organization. Organization theory has received contributions mainly from the disciplines of social sciences, including theoretical and empirical results of economics, political science, sociology and social psychology. Engineering and computer science have provided organizations with mathematical and computational tools to support the activities of analysis and design, and also to explore and to exploit organization theories.

The literature has provided organizations with divergent, complementary and common definitions. Diversity exists because of the different types and shapes of organizations; because of the various backgrounds of the researchers; and due to the different levels of analysis which can focus on social psychological, structural and institutional organizational issues. However, the concepts of rational, natural and open systems have provided organization theory with more precise perspectives on the borders that separate the various definitions of organizations. These concepts together represent a general framework for the analysis of organizations; they also contribute for the classification of organization theories into the perspectives of rational, natural and open systems.

The distinction between formal and informal organizations is just a question of degree, and thus the transition from one to another type is gradual, rather than abrupt.

APPENDIX C. DISCIPLINES OF ORGANIZATION THEORY

C.1. Introduction

Organization theory is a multi-disciplinary field. Most of the contributions it has received are borne from the disciplines of social sciences. This Appendix represents a summary of the main body of disciplines which have supported organization theory.

C.2. Economics

Economics is concerned with economic decision-making, and thus with the efficient allocation of relatively scarce resources to satisfy the needs of human beings. It involves principles of maximization of outcomes through a minimal input of resources. Nevertheless, efficiency of outcomes (and allocation) is not the main concept for the study of economic organizations. Instead, economists are also concerned about the efficiency of the organizations themselves. So, the discipline of economics provides organization theory with approaches to the analyses of different arrangements and forms of organizations, and also to the design of organizations which pursue economic goals.

C.2.1. Organizations as the whole Economy

Economists have defined organizations as created entities within and through which people interact to reach individual and collective economic goals [Milgrom and Roberts, 1992]. At the highest level of economic analysis, organizations are represented by the whole economy, which consists of networks of people, organizations, institutions and regulative processes with the market, and general transactions.

C.2.2. Agency Theory and Organizations

At the next level, which concerns principles of agency theory [Clark, 2000], the organization is viewed as a nexus of contracts among individuals who belong to the classes of principal and agent. Principal is equivalent to owner, and agent is the person or worker who acts on behalf of the principal. At this level, organizations are regarded as formal entities having independent legal identity, which enables them to enter biding contracts, and thus to seek court enforcement of those contracts. Owners seek to maximize their return on investment by the most efficient use of the organization (including the workers). Agents, on the other hand, seek to minimize their efforts and maximize their remuneration. To protect their interests, principals will use various forms of contracts and organizing to ensure that agents carry out their jobs.

C.2.3. Transaction Costs Theory and Organizations

Another level of analysis in economic organization theory is the transaction - i.e. the transfer of goods or services from one individual to another [Milgrom and Roberts, 1992; and Williamson and Masten, 1999]. Such transactions involve costs that take place inside and outside the organization. Such costs of running the organization are mainly originated by the activities of coordination of markets and motivation of people - including transactions between owners and managers, managers and subordinates, suppliers and producers, and sellers and buyers. Transaction costs depend on the nature of the transaction and on the way it is organised. Therefore, organization design at this level is concerned with the choice of organizational structures which better minimizes on these transaction costs.

Both agency and transaction cost perspectives view the primary reason for organizing as being the reduction of uncertainty that exists in typical transactions. Economists have also provided organizations with theories supported by analytical approaches, mainly with contributions received from the discipline of game theory [Gul, 1997]. Table C.1 summarizes such organizational economic perspectives.

Table C.1. Organization Economic Theories

Level of Analysis	Components	Perspectives
Economics	The whole economy	Organizations are created entities within and through which people interact to reach individual and collective economic goals.
Agency Theory	Principal and Agent	Organizations are nexuses of contracts.
Transaction Costs	Firms and markets.	Organizations are firms that coordinate costs associated with internal and external exchanges.

C.3. Political Science

Political scientists have tailored general principles of political science for organizations.

The chief concern of political scientists is the study of power, politics and political parties within the whole society and its derivatives - like government, institutions and unions [Khandwalla, 1977]. In its short definition, power is the ability of agents to allocate and control resources to the ends they favour, by legitimacy, i.e. legitimate power, by charisma, or by force [Scott, 1998]. Politics involve the study of strategies employed by agents - individuals and organizations - to the pursuit of power. Political party means the political organization of a body of people under a governmental system which reflects an ideology, a set of values and beliefs about what the governmental system ought to be.

Political science has been primarily studied through philosophical, institutional and behavioural approaches.

C.3.1. Philosophical Approach

Political philosophy emerged with ancient civilizations, and it received new contributions passing through the periods of the Middle Ages, Renaissance, Enlightenment, and Industrial Revolution up to now. Aristotle (384-322 B.C.), Socrates (469-399 B.C.) and Plato (427-347 B.C.) were distinguished pioneer philosophers with political contributions during ancient civilizations, and also in more recent centuries were Niccolò Machiavelli (1469-1527), Thomas More (1478-1535), Thomas Hobbes (1588-1679), John Locke (1634-1704), Jean-Jacques Rousseau (1712-1788), Thomas Paine (1737-1809), Karl Marx (1818-1883), Frederick Engels (1820-1895), Mao Tse Tung (1893-1976) and Rosa Luxemburg (1871-1919) - among others [Delouche, 2001]. Among their contributions are the studies of alternative forms of government and organizing a society; the origin of a state and the genesis of regulative processes; and the implications of different forms of politics for a state and society.

C.3.2. Institutional Approach

Institutional approaches to political science were dominant in Europe and in United Sates of America during the second half of the 19th century and the first half of the 20th century [Scott, 2001]. Institutional analysis was based in constitutional law, bureaucracies and moral philosophy. Particular attention was given to the study of governmental structures, their functioning and the normative processes within them.

C.3.3. Contemporary Approach

Contemporary political science has advanced from the study of political philosophy and institutional aspects of government to political economy and behaviour. The latter involves the analysis and design of economic political systems, and also the studies of conflicts of interest among agents, relationships between normative and behavioural structures of political systems - including organizations and general political parties - and the influence of normative structures on the behaviour of citizens, and vice-versa - i.e. the influence of citizens on the normative structure of a democratic state by voting for instance. Such studies have given genesis to the research of other subjects related to organizations such as individual (personality, attitudes, perception, learning and motivation), interpersonal (group and team behaviour, conflict and negotiation, leadership and communication) and organizational (decision-making and design, change and culture) processes [Hellriegel *et al*, 2001]. This has led political scientists to integrate sociological, anthropological and psychological concepts such as hierarchy and bureaucracy, culture and socialisation into their background.

C.3.4. Political Science and Organizations

Organizations comprise politics and political parties - i.e. authority systems - and they hold distributed power among agents within their hierarchical structure. Organizations influence the environment through power, and the environment also influences them by the same means, i.e. power. Agents within the organization may have power because of their ability to reward and to punish others; because of their legitimate position in the organizational hierarchy; because of their charisma and social relation with others; or because they posses more knowledge than others on the domains of both the organization and its environment. Moreover, as important as knowledge, are the special cognitive skills that agents have for decision-making and problem solving. The latter ability, on knowledge and cognition, plays the most important role in organizations of today.

Concepts of power, authority, legitimacy, hierarchy of authority and bureaucracy - among others - were firstly introduced into the context of the structure of organizations by Max Weber (1864-1920) [Scott, 1998]. Weber qualified in law and then he became a member of the staff of Berlin University. He had also interest in the broad fields of philosophy, historical development of civilizations, sociology of religion and economy. Nevertheless, those of his writings which have been translated into English have established him as a major contributor to organizational sociology [Pugh and Hickson, 1997]. Table C.2 summarises the tailoring of political analysis to organizations.

Table C.2. Political Analysis of Organizations

Topics of Analysis	Tailoring to Organizations
Power	Analysis of sources of power, distribution of authority, and the implications of power on individual and organizational behaviour.
Politics	Design of strategies and tactics to the achievement of organizational goals, and to the motivation of participants with inducements.
Political Parties	Design of normative structures and analysis of their implications on individual and organizational behaviour.

C.4. Sociology

Sociology concerns the study of systems of social actions, social phenomena and social life. Social action is non-instinctual human behaviour to satisfy needs. Social phenomena emerge from human behaviour and interrelations existing within a social system. The subject of social life ranges - to name but a few of them - from the intimate family to the hostile mob; from organized crime to religious cults; from the divisions of race, gender, wealth categories and social status to the shared beliefs of a common culture; and from the sociology of work to the sociology of organizations. Sociologists investigate the structure of groups, organizations and institutions, communities and societies, and how people interact within these contexts. In fact, few fields have such broad scope and relevance for research, theory and application of knowledge [Khandwalla, 1977].

Organizations resemble miniature societies, and thus they are strong candidates for sociological studies [Blau, 1974; Etzioni, 1969; Grusky and Miller, 1981; Perrow, 1974; and Silverman, 1976]. Researchers have provided the literature with sociological perspectives and social processes of prominent contributions to the analysis and design of organizations [Khandwalla, 1977].

C.4.1. Sociological Perspectives

Sociological perspectives encompass the subjects of stability and order, disorder, continuity and change.

- *Social Stability*

Social stability and order are supported by the equilibrium model of society which resembles the concepts of homeostasis and self-regulation. In organizations, stability and order can assume various definitions, and they can also be achieved by different means. Nevertheless, stability and order can be synonymous of equilibrium, and thus of a theory on organizational equilibrium [March and Simon, 1958]. Organizational equilibrium is essentially a theory of motivation which concerns the alignment of organizational goals with participants' motives. Organizational goals are expected to govern the decisions of the participants in the organization, and motives are individual goals (or expectations) which govern the participants' decision to join and to remain in the organization [Scott, 1998]. Such an alignment can be satisfied, or maximized, by designing inducements - i.e. incentive and reward systems - to the organization [Dunnette and Hough, 1992: 1009-1055]. In such a field, agency theory plays an important role to support economic analysis of organizations [Gibbons, 1998]. Nevertheless, one may needs to design a control system within the

165

organization in order to assure such an alignment [Anthony *et al*, 1984].

- *Social disorder*

Social disorder is supported by explanations of the conflict model of society. Disorder can be synonymous with lack of harmony and organization among different segments of a society. Within organizations, disorder can emerge from diverse factors like from dispute of power; from disagreements of agents with organizational normative structures; from distortions between agents' motives and organizational goals; from the lack of resources; from the lack or inefficiency of incentive and reward systems; from intra-individual and group conflicts [March and Simon, 1993]; and from conflicts between the organization and its environment - which can include technical (e.g. technological requirements and standards), managerial (e.g. coordination of buyers and suppliers), institutional (e.g. regulative systems within the market) and environmental (e.g. natural resources) levels of conflicts. In these cases, sociologists might assume that such phenomena of disorder arise because agents have a high need for power, differentiation in perceptions and limits of cognition. Hence, the roles, the hierarchy of authority, the communication systems, and the strategic reward systems within the organization should be reviewed and redesigned to provide it with stability. In such a way, sociologists avoid to focus on the particularities of the personality of the agents. Nevertheless, they do not necessarily ignore the nature of personality and human needs.

- *Continuity*

Continuity refers to the study of survival of a society and social patterns of behaviour within this society over time, despite the turnover of its members. It can be argued that a society persists and continues to exist with its social life when it is supported by structures which represent the means whereby it satisfies the needs for its survival. In such a way, Parsons developed an approach which identifies the primary functions that all social systems must perform if they are to survive and to persist [Parsons, 1960]. Such functions defined by him are: adaptation (the search and acquisition of resources); goal attainment (the setting and achievement of goals); integration (cooperation and coordination among the parts of a system); and latency (the creation, preservation and transmission of values and culture of a system). Similarly to societies, organizations also have functional needs and structures. However, more than many other types of social structures, organizations are designed to persist over time by supporting complex tasks which go beyond the limits of individual agents. Organizations, when well designed, possess robust structures with capabilities to adapt to both internal and external demands; with abilities to accommodate and to coordinate a broad diversification of activities; with ability to motivate their participants; with flexibility to change their members, technologies, structural features, processes and even goals; and with agility to respond to their environment. In such a way, stability and order play a fundamental role in the survival of organizations, while disorder, viewed as pathology, can lead organizations to instability and extinction.

- *Change*

Change concerns the broad study of social evolution and it has received contributions mainly from Darwinian theories of natural selection, evolution and processes of evolution [Huxley, 1958 and 1974]. Social evolution has inspired researchers to the study of evolution of organizations - i.e. the study on how organizations evolve and adapt to their environment [Cummings and Staw, 1990; and Singh, 1990] - and also on the emergence of new organizational forms among a population of other organizations [Scott, 1998]. In such a way change plays a fundamental role to the continuity and survival of organizations.

C.4.2. Social Processes

Social processes involve a series of activities within a society that when performed lead one to the achievement of ends. Some examples of social processes include classification or stratification of people into status and wealth categories, and deviant behaviour such as crime. In organizations, stratification can be synonymous of division of labour and specialization; departmentalization; delegation of authority or distribution of power; centralization of decisions; promotion of agents according to their qualifications and skills; and so on. Deviant behaviour is likely to take the forms of strikes, conflicts and general disorder. In order to survive, evolve and develop, organizational processes also have to be continuously analysed, designed and redesigned. If structure is thought of as the anatomy of the organization, processes are its physiology or functioning, and thus information and decision processes cut across the organizational structure [Galbraith, 2002]. Organizational processes can be classified as having two foci. The first is concerned with the allocation of scarce resources - like funds and talents - and they encompass activities of planning and budgeting for instance. The second refers to work flows within the activities of requirement management, project and product development - among others.

C.4.3. Sociology and Organizations

Sociology plays an important role in the process of designing organizations by providing people with means (knowledge and technology) to select a candidate among a set of alternatives of organizational forms which better satisfy specific criteria. What is remarkable in the study of organizational sociology is the emphasis on uncovering the structural factors shaping the functioning of organizations. It is because of the sociological preoccupation with structural explanations and with the study of the whole organization, rather than of just groups and individuals within it, that sociologists have sought to explain the structure and functioning of organizations in terms of their elements (like environment, social structure, technology, participants and goals), and also considering their size, shape, age and purpose for instance [Scott, 1998].

Table C.3 summarises the tailoring of sociological perspectives and social processes to organizations.

Table C.3. Sociological Contributions to Organizations

Sociological Perspectives	Application to Organizations
Stability and Order	Design of inducements, and thus incentive and reward systems; alignment of individual motives and organizational goals.
Disorder	Analysis of conflicts and organizational redesign; design of communication systems to equalise individual perceptions.
Continuity	Design of robust organizational forms.
Change	Evolution, adaptation to the environment, and redesign of organizations.
Social Processes	Application to Organizations
Stratification	Analysis of specialization, departmentalization, and delegation of authority.
Deviant behaviour	Analysis of conflicts and general disorder.

C.5. Social Psychology

As derived from the Latin, *psyche* means mind, and *ology* means study, and thus psychology is the science of the mind which studies human behaviour and thinking. The mind involves mental processes, functions and issues like emotions, sensations, thoughts, dreams, consciousness, imagination, perception (among others), which cannot be observed through the physical senses of seeing, hearing, smelling, tasting and feeling. Hence, psychologists have opened three main directions of research called introspection, behaviourism and cognitivism [Reisberg, 1997]. Although such schools share similar goals, i.e. the study of human behaviour and mental processes, they differ in the strategy or approach to the achievement of such goals.

C.5.1. From Introspection and Behaviourism to Cognition

The introspection school was intensive in the late 19th century and it was aimed to explain mental concepts as perception, emotion and consciousness by examining one's own thoughts.

The behaviourist school emerged in the early 20th century as an alternative approach to the lack of scientific basis within the introspection research. Behaviourism is concerned with scientific methods which pursue to provide researchers with the analysis of objective data, i.e. information that can be observed or measured by others. Hence, behaviourism was oriented to stimulus-response and reward theories of human behaviour, learning and intelligence. Nevertheless, although providing the literature with approaches to deal mainly with empirical and observable aspects of human behaviour, the behaviourism school underestimate the study of human mental processes like perception. This was the point of departure for the emergence of a new direction of research called cognitive psychology.

The transition from behaviourism to cognitivism was paved by the school of gestalt psychology in the first half of the 20th century. Gestalt scientists were extensively concerned with the study of human perception through the use of animals (as well as humans). Cognitive psychology is oriented to the study of high mental processes which govern human behaviour, learning and intelligence, and thus it investigates theoretical models and those concepts underlying the functioning of the human mind, like perception, attention, categorization, knowledge representation and organization, memory, language, decision-making and problem solving. However, greater advances in cognitive research were paved by the development of the digital computer and the discipline of artificial intelligence around the 1950's. These gave power to the emergence of new theories and models of the human mind. Among these models are information-processing systems [Newell and Simon, 1972] and others on multiple human minds as distributed computational agents [Minsky, 1986; and Carley and Gasser, 1999]

The schools of introspection, behavioural and cognitive psychology have also provided new insights to the research and understand of human evolution and development [Butterworth *et al*, 1985; and Heyes and Huber, 2000]. Lefrançois provides a complete overview about these schools, and he also presents a clear analysis on the transition and re-orientation from introspection and behaviourism to cognitivism [Lefrançois, 1995].

C.5.2. Psychology and Its Social Context

Social psychology is a branch of psychology which comprises introspective, behaviourist and cognitive theories, but extended to individuals in a social context. Social psychology investigates those phenomena which emerge within interpersonal relations. It focuses on how human behaviour and mental activities - and thus personality, attitudes, motivation, emotions, beliefs and also perception, attention, learning, decision-making and problem solving processes of individuals - are influenced by other people, technologies, machines, groups and teams, organizations, normative and regulative processes, and by any other social factor.

Social psychology includes different and complementary orientations like gestalt, field theory, reinforcement, psychoanalytic and role theory [Khandwalla, 1977].

- *Gestalt Psychology*

Put shortly, the gestalt orientation provides analysis on how individuals' perception, beliefs and cognitive processes are influenced by the presence of others.

- *Field Theory Orientation*

Field theory, as influenced by gestalt conceptions, has been concerned with the dynamics of human motivation, and thus needs and goals. It asserts that people's motivations, attitudes, opinion and hence behaviour can change due to the influence of others. Hence, field theory provides analyses on conflict, cooperation and coalition among agents and groups.

- *Reinforcement Learning*

Reinforcement is a branch of learning theories and it is synonymous of behavioural change. It asserts that behaviour is a product of conditioning imposed by external forces. In such a way humans are viewed as passive instruments whose behaviour is regulated by the environment. The main principles of regulation of behaviour are those of rewarding and punishing, and also the frequency of reinforcements.

- *Psychoanalytic Research*

The psychoanalytic orientation received most of its contribution from Sigmund Freud (1856-1939). Psychoanalysis has provided social psychologists with studies on personality formation and change, and also on authoritarian and pathological behaviour (including stress, anxiety, frustration, conflict and tension).

- *Role Theory*

If personality is synonymous with similarities in the responses of an individual to different situations, role represents the similarities in the responses of different individuals to a same situation. So, in a given culture, a role is synonymous with prescribed behaviour associated with a given status or position. Role theory investigates the influence of roles on shaping personality and vice-versa, and also role conflict and deviant behaviour. Role conflict can arise because of incompatibilities between the personality of an individual and the roles he has to fulfil, and also because of incompatible and high demands of different roles on the same individual. Deviant behaviour emerges when individuals do not play their appointed roles.

C.5.3. Social Psychology and Organizations

In the context of social psychology, organizations are viewed as arenas and networks relationships with interlinked mental processes and behavioural phenomena which shape their structure, functioning and also the behaviour of their participants. Therefore, social psychology plays an important part into the domain of organization behaviour and management science [Carley and Gasser, 1999; and Hellriegel *et al*, 2001].

Table C.4 presents a summary on the perspectives of social psychology within the context of organizations, and on how such perspectives are viewed in organizational analysis and design.

While sociologists place emphasis on structural and functional conditions which shape events within organizations (similarly to a top-down approach moving from macro to micro level of analysis), social psychologists try to explain and predict macro-level behaviour, such as overall organizational performance, from the analysis of micro-level behaviour which emerges from complex interactions among the participants in the organization, towards a bottom-up approach.

Table C.4. Social Psychology Perspectives on Organization: Analysis and Design

Perspectives	Analysis	Design Issues
Social Psychology (Gestalt, field theory, reinforcement, psychoanalytic, and role theory orientations)	Organizations are viewed as arenas of interpersonal relations with interlinked mental processes and behavioural phenomena.	Alignment of: normative and behavioural structures; organizational goals and individuals' motives; roles and personalities. Use of reward and control systems. Incentive to communication and problem solving.

C.6. Engineering

Engineering has contributed to the field of organizations with disciplines and methodologies of systems analysis and design. Among such disciplines are industrial and systems engineering, operational research, and engineering economics and management.

Industrial and systems engineering, and operations research, traditionally focus on the analysis and design of manufacturing and service systems for the efficient production and distribution of goods and services. Engineering economics concerns the application of principles of economy and engineering to the analysis and design of systems. Therefore, it involves trade-offs between economical and technical decisions. The role of engineering management is to bridge the gap between engineering and management activities. It provides engineers with principles of organizations and management. On the other hand, it provides managers with engineering tools which pursue the optimization of managerial and organizational processes through the use of scientific methods, like those within operational research and artificial intelligence. Therefore, it is also synonymous of industrial engineering. Table C.5 summarises the contributions of engineering to organizations.

Table C.5. Engineering and Organizations

Subjects	Concerns
Industrial and Systems Engineering, and Operational Research	Analysis and design of manufacturing and service systems, and their integration to other parts. They provide tools for searching optimal and efficient solutions.
Engineering Economics	Economical and technical decision analysis of projects and organizational forms.
Engineering Management	Fulfil the gap between technical and managerial tasks. It provide tools for optimizing managerial processes like operational research.

C.7. Computer Science

Computer science has provided organization theorists with new principles and tools to the analysis and design of organizations. Computational modelling for instance has provided researchers with models of organizations which are synonymous of theories [Cohen and Cyert, 1965]. By simulating such models in computers, theories can be explained and tested, organizational behaviour can be predicted under new assumptions, and additional propositions can be derived to constitute new theories. Moreover, computational simulations can be used to support organizational design by determining for instance which among several alternatives of organizational forms are best suited to satisfy specific goals as settled to the organization.

Computer science has also provided new approaches to the analysis of organizations by resembling them to the principles of information processing systems, distributed computational agents [Blanning and King, 1996; and Carley and Gasser, 1999] and artificial life [Langton, 1995]. However, it shares such contributions mainly with the disciplines of psychology, artificial intelligence and biology. Table C.6 summarises the contributions of computer science to organizations.

Table C.6. Computer Science and Organizations

Subjects	Concerns
Computational Modelling	Design of organizational models.
Computational Simulation	Analysis of organizational forms; analysis of organizational behaviour, decisions and theories; it provides evidence for theories and it makes possible the proposal of new theories from massive data analysis.
Computational Organizational Theory (COT)	A new discipline for modelling and simulating organizations as distributed computational agents. It has been largely studied at the Carnegie Mellon University. Section A.4 describes COT in *Computational Modelling and Simulation*.

APPENDIX D. TECHNOLOGY

D.1. Introduction

This Appendix surveys the subject of technology. Furthermore, it fits technology into the domain of organizations. It starts by presenting a perspective on the genesis of technology, and by proposing a definition of technology. It explains the benefits of technology to organizations, and it also proposes some rationales for technology into a political, economic and social context. It explains the scope of technology in organizations, which ranges from different levels of analysis to distinct elements of application. Additionally, it explains why *cognitive machines* represent a challenge for scientists of organizations and technology. This Appendix concludes by proposing that the Industrial Revolution has already turned its emphasis from energy to information-demanding technologies.

D.2. A Perspective on the Genesis of Technology

Humans have needs, but even when they satisfy their needs, they want more and more. Perhaps at first it was to obtain more food, or to find better shelter. Nowadays, however, the desires of humans seem endless. People want to cure diseases, to explore continents, the sea and the universe, to reshape nature and to build artificial habitats like cities and airports, to replace muscular and cognitive tasks with machines, and even to live longer and longer [Kipnis, 1990].

Most of human needs are created by conditions imposed by the environment on them. They answer by imposing changes on the environment. Most animals survive by adapting to the environment. However, although humans do the same, they also survive by adapting the environment to themselves.

The genesis of technology resides in the human ability to search knowledge. Humans transform knowledge in identifiable ends. Ends have economic, social and political facets, and they can be ends by themselves or even other means used to achieve more complex goals. Moving upwards from means to ends there exist sub-goals, goals and more complex goals at upper levels. Therefore, technology can also be conceived as processes of hierarchical order and control.

D.3. A Definition of Technology

Technology is a broad discipline which has received diverse definitions in the literature. However, all of them include knowledge as a common precept [Anthony and Gales, 2003; Goodman, 1990; Richter, 1982; Scott, 1998; and Simon, 1977]. Similarly to the definition of organizations, the conception of technology within this section avoids any divergence from the literature. Moreover, it represents a synthesis aligned to this research scope.

Technology represents means deliberately employed by humans for attaining practical outcomes. Technology encompasses knowledge, and furthermore, processes to produce, to process and to manage knowledge. It also includes tools, practices and even other technologies.

In such a way, science plays an important part in the conception of technology, since it is defined as the use of scientific methods to search knowledge. Science can be classified in natural and artificial forms [Simon, 1996]. Put shortly, the former is knowledge about natural objects and phenomena, and it is concerned with analysis and discoveries - like the Newton's law of gravitation for instance. The latter is knowledge about artificial (and synthetic) objects and phenomena; it is concerned with engineering and inventions, man-made things, and thus with design - like knowledge on agriculture, medicine, machines and organizations for instance.

D.4. Benefits of Technology to Organizations

Technology reduces the amount of uncertainty in the attempt to solve practical problems, and it increases the probability of occurrence of events as wanted by power-holders [Kipnis, 1990; and Perrow, 1967].

This research supports this statement and extends it to the next chain of propositions:

Proposition D.4.1: Technology increases the level of complexity of the organization, and it relatively reduces the level of complexity of the environment.

Firstly, proposition C.4.1 assumes by definition that the higher the level of complexity of the organization, the higher its degrees of cognition, intelligence and autonomy. Secondly, it does not mean that the level of complexity of the environment reduces, but it says that such a level of complexity is relatively reduced when compared to the growth in the level of complexity of the organization. Therefore, one can proceed by stating that:

Proposition D.4.2: The higher the level of complexity of the organization, the higher are its degrees of cognition, intelligence and autonomy.

Proposition D.4.3: The higher the degrees of cognition, intelligence and autonomy of the organization, the lower is the relative level of complexity of its environment.

Proposition D.4.4: The lower the relative level of complexity of the environment, the less is the relative amount of uncertainty.

Similarly, proposition D.4.4 says that the amount of uncertainty in the environment is relatively reduced with an increase in the degrees of cognition, intelligence and autonomy of the organization. Therefore, one can deduce the next theorem from the previous chain of propositions:

Theorem D.4.1: Technology increases the level of complexity of the organization, and it relatively reduces the amount of uncertainty of the environment.

Such propositions are further discussed in Chapter 2.

D.5. Rationales for Technology: Political, Economic and Social Contexts

Among the major achievements of humans through technology is the progressive economy of time and energy that one must devote to any given activity, including both physical and mental tasks. By creating technologies of automation, humans can direct their effort upward into more complex tasks.

To think about technology as synonymous with machines and processes only is to reject its broad significance and implications for the society. As organizations do, technology emerges from political, economic and social contexts; moreover, technology also can give reverse implications for such contexts[54]. Consider for instance the expansion of Europe between the 15th and 18th centuries, which was empowered by the dominance of the colonizers in navigation and army technologies. Their overseas discoveries brought about a revolution in the history of humanity, with political, economic and social facets [Delouche, 2001].

D.5.1. Political Context

The most important political facet of technology is power [Kipnis, 1990; and Scarbrough and Corbett, 1992]; technology provides people with it. These people are named power-holders when they control the technology. Power-holders pursue the achievement of complex goals and practical outcomes by exercising control over the environment, and thus by overcoming the resistance of nature and people who are target of control - e.g. workers, consumers, citizens, etc. Therefore, technology can give those who govern the technology control over others, either by taking away some of their alternatives or by constraining them to particular choices. A hierarchy of power-holders is then synonymous with a hierarchical process of decision control.

D.5.2. Economic Context

Technology also emerges from economic reasons and needs, and it reversely feedback a society with new outcomes. The economic facet of technology concerns the efficient production of goods and services. Therefore, technology is supposed to provide people with better standard of life and wealth. However, power-holders are who control the distribution of economic outcomes.

D.5.3. Social Context

On the one hand, some studies on technology have demonstrated that it can provide people with better standard of life and other social outcomes [Easterlin, 2000; and Johnson, 2000]. Consider for instance the advent of communication systems with the radio, the television, the telephone and the internet. With such technologies one can meet (or get in contact with) people from any continent; get more and more knowledge, faster than in any period of the human history; and even work in home with the benefit to spend more time with the family. On the other hand, some authors advocate that technology, and thus the use of power, transforms social relations between the more and the less powerful [Kipnis, 1990]. Other authors (orthodox Marxists) say that technology (and in particular automation) may provide massive unemployment and societal pathologies, like alienation, inequity, social stratification, etc.

This research asserts that new technologies may force people to move from one kind of job to another. The new jobs will be carried out by more complex mental tasks and they will require greater mental efforts than before. However and foremost, unemployment will have to be managed between transitions of technologies. Additionally, institutional processes will have to play a prominent part to regulate and to intermediate consensus between organizations

---

[54] Consider for instance the perspectives on destruction (on world and nuclear wars) and automation (which concerns the efficient and materialist world, where man is eliminated from the economic system), as reviewed in [Rogers, 1975: Chapter 13].

and their employees; and broadly speaking, between organizations and the environment.

D.6. The Scope of Technology in Organizations

Technology is a discipline which encompasses distinct elements of application and different levels of analysis. Elements of technology vary from machines to normative and regulative processes. Similarly to organizations, analysis of technology ranges from technical, managerial and institutional to worldwide systems. These systems concern the technologies of machines and the processes used to carry out activities and cognitive tasks in the respective levels of analysis (or cognitive systems) of the organization. The perspective of the organization as hierarchic cognitive systems along with the organization levels of analysis are presented in the Chapter 2.

Notes: The technical level refers to the set of possible arrangements of machines and processes employed to produce desired outcomes, goods and services. The literature calls it technical system [Scott, 1998]. Furthermore, this level also encompasses the technology of the machines, characterized by their structure (anatomy) and processes (functioning or physiology). This latter refers to the processes embedded in machines and those processes (or protocols) of communication used to connect the machines to a network. Hence, such processes play the preponderant role in the design of *cognitive machines*, by providing them with capabilities to carry out high mental tasks in organizations.

Additionally, such *cognitive machines* are also used to carry out activities and cognitive tasks at the managerial, institutional and worldwide levels. However, as one moves from technical towards worldwide level of analysis, the processes within these machines and systems become more complex. At upward levels, goals and decisions become more complex, uncertainty increases, and the predictability of consequences of choices and outcomes diminishes.

D.7. A Challenge for Today: *Cognitive Machines* in Organizations

Consider for instance a technical system constituted by machines. The interdependence among its parts is such that their behaviour is highly constrained and limited to the prescriptions of a normative structure[55]. This structure is relatively rigid and the system of relations determinant. By contrast, in social systems such as organizations, the connections among the interacting parts are somewhat less constrained, providing them with more flexibility of response. Moreover, organizations have a structure of normative and behavioural parts. The latter part focus on actual behaviour rather than on prescriptions for behaviour, and thus it provides organizations with a more complex behaviour[56]. Therefore, the elements of social systems, and the relationships between them, hold higher levels of complexity than those of technical systems. Social systems such as organizations possess emotions along with higher degrees of cognition, intelligence and autonomy.

---

[55] Such a normative structure of technical systems and machines includes programmed and non-programmed decisions. An overview on programmed and non-programmed decisions is found in [Simon, 1977].

[56] In this particular case complex behaviour is synonymous with unpredictable outcomes.

Nevertheless, artificial systems of high levels of complexity (and thus possessing high degrees of cognition, intelligence and autonomy) have been developed with the advances in technologies of machines. This includes software programs, computers and communication networks, and also developments with the advent of the disciplines of artificial intelligence [Luger and Stubblefield, 1998; and Zadeh, 1999 and 2001] and soft computing [Zadeh, 1994 and 1997]. Such systems have started to play a fundamental and an increasing role in the society of today, by supporting individual and distributed cognitive tasks in organizations [Blanning and King, 1996]. Therefore, the integration of these systems into organizations of today constitutes a subject of challenge for researchers of technology and social sciences.

There is no doubt that technological principles of the past and the present have contributed with brilliantly successful applications in many areas of organizations, such as manufacturing and management [Simon, 1977]. But these successes should not obscure the fact that the world is changing, that high machine intelligence is becoming reality [Zadeh, 1996b], and that methods which have proved to be successful in the past may not provide the right tools for addressing the problems of the future.

D.8. The Industrial Revolution: From Energy to Information

In discussing about technological transformations paved by the Industrial Revolution, it is relevant to distinguish two polar phases in the extremes of this continuous period of the history. These extremes are characterized by energy-based technologies and information-based technologies [Simon, 1977].

At one end, one has energy-based technologies which are characterized by machines with capabilities to replace mainly intensive muscular tasks in organizations.

At the other end, one has information processing-based technologies which are those machines carrying knowledge on how to produce and to manage information more effectively and efficiently. Moreover, they are consumers of a quite modest amount of energy and materials - like modern computers.

Of course one can find technologies of all shades of grey along the continuum between these two extremes of black and white. Machines that autonomously[57] transform one type of energy in another represent the first extreme. Thus, one can assert that they have already been designed since the beginning of the Industrial Revolution in the 18th century.

The other extreme is represented by machines which posses levels of cognition, intelligence and autonomy comparable with humans. These extreme machines have not been completely designed yet, but man has been continuously moving towards the design of them along a continuous path of revolution.

Figure D.1 illustrates two symbolic functions which represent the demands of machines for energy and information, when one moves from the beginning of the Industrial Revolution, up to now, to its continuous development.

Moving along the continuous period of Industrial Revolution, on the horizontal line, it can be observed that information is increasingly demanded, produced and managed by

---

[57] In this context, autonomous means independent of the intervention of human muscles for the transformation of energy. Therefore, old mills - for instance - are not autonomous machines.

machines, and thus it becomes relatively more important than energy for organizations of today. Nevertheless, machines at shop floor still demand a great amount of energy, but not greater than the demand required by those machines of similar purpose as designed in the first phase of the Industrial Revolution.

Therefore, figure D.1 does not mean that energy is being less consumed nowadays in the world than before, but it does assert that nowadays each unit of machine demands less energy and more information[58] than those machines of similar purpose which were designed in the past.

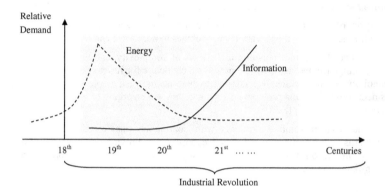

Figure D.1. Relative Demand of Energy and Information over Time

D.9. Summary

Technology encompasses knowledge, and furthermore, tools, practices, processes and other technologies to produce, to process and to manage knowledge.

Concerning about the subject of technology in organizations, this research point of departure is twofold. Firstly, it asserts that organizations expand what people can achieve. Secondly, it states that technology expands what organizations can do. Despite being slightly distinct, such premises complement each other. Organizations and technology are complementary means which extend people capability for the achievement of more complex goals.

Technology arises from the human ability to search knowledge in order to transform it in identifiable ends for their own well-being. It provides humans with means deliberately used for attaining practical outcomes.

Technology can reduce the amount of time and energy which people must devote to any

---

[58] The term "demand of information" means the capability of machines to create and to manage information.

given activity, including both physical and mental tasks.

Technology plays an important part in connecting organizations to the environment. Technology can increase the level of complexity of the organization by improving the organization's powers of cognition, intelligence and autonomy. Hence, technology in one sense relatively reduces the amount of uncertainty in the environment.

Technology emerges from political, economic and social contexts and it provides people with power, economic use of resources and social outcomes such as wealth and well-being. However, it simultaneously provides power-holders with control over the environment. Hence, technology can also give those who govern the technology control over others, either by taking away some of their alternatives or by constraining them to particular choices. These steps form a hierarchical process of decision control.

Analysis of technology ranges from technical, organizational and institutional to worldwide levels, and its elements vary from machines to normative and regulative processes.

The advent of machines pursuing high levels of complexity, and thus high degrees of cognition, intelligence and autonomy, represents a challenge for researchers of organizations and technology. Such machines are coming to play an increasing role in organizations of today as decision-makers and problem-solvers, and they are emerging to work in the name of organizations, just like people do.

The Industrial Revolution proceeds with continuous transformations. It has been moving towards new directions to rationalize energy and to provide man with the design of *cognitive machines*. Such revolution has already shifted its emphasis from energy to information-demanding machines.

APPENDIX E. DEFINITION OF THE PERFORMANCE FACTORS

E.1. Introduction

Appendix C presents definitions of the state variables (X) of the software project process of the Radio Engineering Division of NEC do Brasil S.A. (NDB). These variables are denotations of performance factors of the organization process and they comprise project schedule (T), project cost (C) and product requirements completeness (R).

E.2. Project Schedule (T)

It is concerned with the periods of time or dates associated with the completion of the activities of a project. It is classified into planning and actual schedule.

E.2.1. Planning Schedule (T_o)

It represents the expected period of time or date for the completion of a project and it is estimated during the phase of project planning. T_o may also be defined in terms of the project's milestones or according to other specific granularities of the schedule.

E.2.2. Actual Schedule (T_A)

It represents the actual period of time which was expended for the completion of a project. Similarly to T_o, T_A may also be defined in terms of the project's milestones or according to other specific granularities of the schedule.

E.3. Project Cost (C)

It is concerned with the financial investment put in a project and it is classified into planning and actual costs.

E.3.1. Planning Cost (C_o)

It represents the expected cost of a project and it is estimated during the phase of project planning.

E.3.2. Actual Cost (C_A)

It represents the actual cost of a project computed at time T_o.

E.4. Product Requirements Completeness (R)

It is concerned with measures that indicate the degree of completeness of the requirements of a product as expected or ordered by the customer. It is classified into planning and actual requirements completeness.

E.4.1. Planning Requirements Completeness (R_o)

It represents the total number of requirements of a product as expected or ordered by the costumer.

E.4.2. Actual Requirements Completeness (R_A)

It represents the number (or subset) of requirements of a product which were successfully achieved at time T_o.

APPENDIX F. MENTAL MODELS OF THE *COGNITIVE MACHINE*

F.1. Introduction

This appendix describes the fuzzy conditional statements of the two fuzzy rule bases of the *cognitive machine* designed in the industrial case of Chapter 5. These rule bases are linguistic descriptions of mental models about customer satisfaction (CS) and projects' process quality (PPQ), where CS and PPQ denote performance indexes of the organization of study. Such mental models represent the perceptions and experience of the participants in the Telecommunications Management Networks Section of the Radio Engineering Department of NDB (NEC do Brasil S.A.).

F.2. Linguistic Statements about Customer Satisfaction

(1) IF R is *empty* THEN CS is *very low*

OR

(2) IF R is *almost empty* THEN CS is *low*

OR

(3) IF R is *partial* THEN CS is *medium*

OR

(4) IF R is *almost full* THEN CS is *high*

OR

(5) IF R is *full* THEN CS is *very high*

F.3. Linguistic Statements about Projects' Process Quality

(1) IF C is *cheap* AND R is *empty* THEN PPQ is *bad*

OR

(2) IF C is *cheap* AND R is *almost empty* THEN PPQ is *moderate*

OR

(3) IF C is *cheap* AND R is *partial* THEN PPQ is *good*

OR

(4) IF C is *cheap* AND R is *almost full* THEN PPQ is *very good*

OR

(5) IF C is *cheap* AND R is *full* THEN PPQ is *really good*

OR

(6) IF C is *not so cheap* AND R is *empty* THEN PPQ is *very bad*

OR

(7) IF C is *not so cheap* AND R is *almost empty* THEN PPQ is *bad*
OR

(8) IF C is *not so cheap* AND R is *partial* THEN PPQ is *moderate*
OR

(9) IF C is *not so cheap* AND R is *almost full* THEN PPQ is *good*
OR

(10) IF C is *not so cheap* AND R is *full* THEN PPQ is *very good*
OR

(11) IF C is *expensive* AND R is *empty* THEN PPQ is *really bad*
OR

(12) IF C is *expensive* AND R is *almost empty* THEN PPQ is *very bad*
OR

(13) IF C is *expensive* AND R is *partial* THEN PPQ is *bad*
OR

(14) IF C is *expensive* AND R is *almost full* THEN PPQ is *moderate*
OR

(15) IF C is *expensive* AND R is *full* THEN PPQ is *good*

APPENDIX G. BOUNDARIES AND CONVERGENCE OF THE *COGNITIVE MACHINE*

G.1. Introduction

This appendix is concerned with boundary studies and the proof of convergence of the output variables of the *cognitive machine* designed in Chapter 5. It complements the studies of quantitative analysis by demonstrating Theorem 5.11.2.4.

The state space of the *cognitive machine* designed in the Chapter 5 can be broken down and classified into the following regions or intervals of operation.

G.2. Regions of Operation of the *Cognitive Machine*

G.2.1. State Space of (R → CS)
Region 1: $(0 \leq R_A \leq 0.6Ro)$

Region 2: $(0.6Ro < R_A \leq 0.7Ro)$

Region 3: $(0.7Ro < R_A \leq 0.8Ro)$

Region 4: $(0.8Ro < R_A \leq 0.9Ro)$

Region 5: $(0.9Ro < R_A \leq Ro)$

Region 6: $(Ro < R_A \leq \eta$, for $\eta < \infty)$

G.2.2. State Space of (C AND R→ PPQ)
Region 1: $(0 \leq C_A \leq Co)$ and $(0 \leq R_A \leq 0.6Ro)$

Region 2: $(0 \leq C_A \leq Co)$ and $(0.6Ro < R_A \leq 0.7Ro)$

Region 3: $(0 \leq C_A \leq Co)$ and $(0.7Ro < R_A \leq 0.8Ro)$

Region 4: $(0 \leq C_A \leq Co)$ and $(0.8Ro < R_A \leq 0.9Ro)$

Region 5: $(0 \leq C_A \leq Co)$ and $(0.9Ro < R_A \leq Ro)$

Region 6: $(0 \leq C_A \leq Co)$ and $(Ro < R_A \leq \eta)$

Region 7: $(Co < C_A \leq 1.25Co)$ and $(0 \leq R_A \leq 0.6Ro)$

Region 8: $(Co < C_A \leq 1.25Co)$ and $(0.6Ro < R_A \leq 0.7Ro)$

Region 9: $(Co < C_A \leq 1.25Co)$ and $(0.7Ro < R_A \leq 0.8Ro)$

Region 10: $(Co < C_A \leq 1.25Co)$ and $(0.8Ro < R_A \leq 0.9Ro)$

Region 11: $(Co < C_A \leq 1.25Co)$ and $(0.9Ro < R_A \leq Ro)$

Region 12: $(Co < C_A \leq 1.25Co)$ and $(Ro < R_A \leq \eta)$

Region 13: $(1.25Co < C_A \leq 1.5Co)$ and $(0 \leq R_A \leq 0.6Co)$

Region 14: $(1.25Co < C_A \leq 1.5Co)$ and $(0.6Ro < R_A \leq 0.7Ro)$

Region 15: $(1.25Co < C_A \leq 1.5Co)$ and $(0.7Ro < C_A \leq 0.8Ro)$

Region 16: $(1.25\text{Co} < C_A \le 1.5\text{Co})$ and $(0.8\text{Ro} < C_A \le 0.9\text{Ro})$

Region 17: $(1.25\text{Co} < C_A \le 1.5\text{Co})$ and $(0.9\text{Ro} < C_A \le \text{Ro})$

Region 18: $(1.25\text{Co} < C_A \le 1.5\text{Co})$ and $(\text{Ro} < R_A \le \eta)$

Region 19: $(1.5\text{Co} < C_A \le \eta)$ and $(0 \le R_A \le 0.6\text{Co})$

Region 20: $(1.5\text{Co} < C_A \le \eta)$ and $(0.6\text{Ro} < R_A \le 0.7\text{Ro})$

Region 21: $(1.5\text{Co} < C_A \le \eta)$ and $(0.7\text{Ro} < C_A \le 0.8\text{Ro})$

Region 22: $(1.5\text{Co} < C_A \le \eta)$ and $(0.8\text{Ro} < C_A \le 0.9\text{Ro})$

Region 23: $(1.5\text{Co} < C_A \le \eta)$ and $(0.9\text{Ro} < C_A \le \text{Ro})$

Region 24: $(1.5\text{Co} < C_A \le \eta)$ and $(\text{Ro} < R_A \le \eta)$, for $\eta < \infty$

G.3. Mathematical Analysis

The regions of operation of both CS and PPQ can be described mathematically by the application of the results of Theorems 5.11.2.2 and 5.11.2.3. In the following, Theorem 5.11.2.2 is applied to describe the analytical equation that describes the region of operation 2 of customer satisfaction (CS): $(0.6\text{Ro} < R_A \le 0.7\text{Ro})$.

$$CS(t) = -\frac{10}{R_o}(R_A - 0.7R_o).0 + \frac{10}{R_o}(R_A - 0.6R_o).2.5 \qquad \text{(G.1)}$$

$$CS(t) = \frac{25}{R_o}(R_A - 0.6R_o) \qquad \text{(G.2)}$$

By calculating the limits of $CS(t)$ in the boundaries of the region 2, it is found that:

$$\lim_{R_A \to 0.6R_o} CS(t) = 0 \qquad \text{(G.3)}$$

$$\lim_{R_A \to 0.7R_o} CS(t) = 2.5 \qquad \text{(G.4)}$$

It can be concluded that the value of CS in the region of operation 2, as modelled by the equation G.2, converges to a bounded real value in the interval [0,10] as the input variables C and R tend to the limits of operation defined to this region. The same results can be concluded to the overall regions of operation of CS and PPQ. Therefore, the output variables of the *cognitive machine* are bounded and they converge to real values in the interval [0,10] defined to their respective universes of discourse.

———

REFERENCES

A

[Alavi, M. and Palmer, J. 2000] Use Information Technology as a Catalyst for Organizational Change. In *The Blackwell Handbook of Principles of Organizational Behavior*, ed. by Locke, E.A. 2000: 404-415. Blackwell Publishers Ltd.

[ANATEL] National Agency of Telecommunications of Brazil: http://www.anatel.gov.br

[Anthony, R.N., Dearden, J. and Bedford, N.M. 1984] *Management Control Systems*. Richard D. Irwin, Inc.

[Argote, L. 1999] *Organizational Learning: Creating, Retaining and Transferring Knowledge*. Kluwer Academic Publishers.

[Argyris, C. and Schön, D.A. 1978] *Organizational Learning: A Theory of Action Perspective*. Addison-Wesley.

[Ashby, W.R. 1968] Principles of the Self-Organizing System. In *Modern Systems Research for the Behavioral Scientist*, ed. by Buckley, W. 1968: 108-118. Aldine Publishing Company.

[Augier, M. and March, J.G. 2002] *The Economics of Choice, Change and Organization: Essays in Memory of Richard M. Cyert*. Edward Elgar,

[Autor, D.H. 2001] Wiring the Labor Market. *The Journal of Economic Perspectives*, 15 (1): 25-40.

[Axelrod, R.M. 1997] *The Complexity of Cooperation: Agent-Based Models of Competition and Collaboration*. Princeton University Press.

B

[Bäck, T., Fogel, D.B. and Michalewicz, Z. 2000] *Evolutionary Computation: Part I and II*. Institute of Physics Publishing.

[Bagozzi, R.P., et al. 1998] Goal-directed Emotions. *Cognition and Emotion*, 12 (1): 1-26.

[Bahg, C. 1990] Major Systems Theories Throughout the World. *Behavioral Science*, 35 (2): 79-107.

[Bailey, K.D. 1982] *Methods of Social Research*. The Free Press.

[Bakos, Y. 2001] The Emerging Landscape for Retail E-Commerce. *The Journal of Economic Perspectives*, 15 (1): 69-80.

[Barber, B.M. and Odean T., 2001] The Internet and The Investor. *The Journal of Economic Perspectives*, 15 (1): 41-54.

[Barnard, C.I. 1938] *The Functions of the Executive*. Cambridge, Mass.

[Bernstein, D.A. et al 1997] *Psychology*. Houghton Mifflin Company.

[Barsalou, L.W. 1999] Perceptual symbol systems. *Behavioral and Brain Science*, 22: 577-660.

[Bertalanffy, L. von, 1962] General System Theory: A Critical Review. *General Systems*, VII: 1-20.

[Bertalanffy, L. von, 1968] *General system theory: foundations, development, and applications*. Allen Lane.

[Black, M. 1937] Vagueness: An Exercise to Logical Analysis. *Philosophy of Science*, 4: 427-455.

[Black, M. 1963] Reasoning with Loose Concepts. *Dialogue*, 2:1-12.

[Blanning, R.W. and King, R.K. 1996] *AI in Organizational Design, Modeling, and Control*. IEEE Computer Society Press.

[Blau, P.M. 1974] *On the Nature of Organizations*. John Wiley & Sons.

[Blau, P.M. 1981] The Comparative Study of Organizations. In *The Sociology of Organizations: Basic*

Studies, ed. by Grusky, O. and Miller, G. (1981): 110:128. The Free Press.

[Blau, P.M. and Scott, W.R. 1963] *Formal Organizations: A Comparative Approach*. Routledge.

[Boland, R.J., and Tenkasi, R.V. and Te'Eni, D. 1996] Designing Information Technology to Support Distributed Cognition. In *Cognition Within and Between Organizations*, ed. by Meindl, J.R., Stubbart, C. and Porac, J.F. (1996): 245-280. Sage Publications.

[Bond, A.H. and Gasser, L. 1988] *Readings in Distributed Artificial Intelligence*. Morgan Kaufmann Publishers, Inc.

[Borenstein, S. and Saloner, G. 2001] Economics and Electronic Commerce. *The Journal of Economic Perspectives*, 15 (1): 3-12.

[Boulding, K.E. 1956] General Systems Theory: The Skeleton of Science. *Management Science*, 2: 197-208.

[Boulding, K.E. 1966] *The Impact of the Social Sciences*. Rutgers University Press.

[Boulding, K.E. 1978] *Ecodynamics: A New Theory of Societal. Evolution*. SAGE Publications.

[Bowman, E.H. and Kogut, B.M. 1995] *Redesigning the Firm*. Oxford University Press.

[Breazeal, C. 2000] Sociable Machines: Expressive Social Exchange between Humans and Robots. *Sc.D. Dissertation*, Department of Electrical Engineering and Computer Science, MIT.

[Brown, S.A. 2000] *Customer Relationship Management: Linking People, Process and Technology*. John Wiley Trade.

[Brynjolfsson, E. and Hitt, L.M. 2000] Beyond Computation: Information Technology, Organizational Transformation and Business Performance. *The Journal of Economic Perspectives*, 14 (4): 23-48.

[Buckley, J. and Ying, H., 1989] Fuzzy Controller Theory: limit theorems for linear fuzzy control rules. *Automatica*, vol. 25, (3): 469-472.

[Buckley, W. 1968] *Modern Systems Research for the Behavioral Scientist*. Aldine Publishing Company.

[Bunge, M. 1979] *Treatise on Basic Philosophy. Ontology II: A World of Systems*. D. Reidel Publishing Company.

[Bunge, M. and Ardila, R. 1987] *Philosophy of Psychology*. Springer-Verlag.

[Butterworth, G., Rutkowska, J. and Scaife, M. 1985] *Evolution and Developmental Psychology*. The Harverster Press.

C

[Carley, K.M. and Gasser, L. 1999] Computational Organizational Theory. In *Multiagent Systems: A Modern Approach to Distributed Artificial Intelligence*. Ed. by Weiss, G. (1999: 299-330). The MIT Press.

[Caplow, T. 1964] *Principles of Organization*. Harcourt, Brace & World, Inc.

[Clark, P.A. 2000] *Organizations in action: competition between contexts*. Routledge.

[Clark, P.A. 2003] *Organizational Innovations*. SAGE Publications.

[Cohen, K.J. and Cyert, R.M. 1965] Simulation of Organizational Behavior. In *Handbook of Organizations*, ed. by March, J.G. 1965: 305-334. Rand McNally & Company.

[Cummings, L. and Staw, B. 1990] *The Evolution and Adaptation of Organizations*. Jai Press Inc.

[Cyert, R.M. and March, J.G. 1963] *A Behavioral Theory of the Firm*. 1st Ed. Blackwell Publishers.

D

[Daft, R.L. 2001] *Organization Theory and Design*. South-Western College Publishing.

[Daft, R.L. and Noe, R.A. 2001] *Organizational Behavior*. Harcourt, Inc.

[Delouche, F. 2001] *Illustrated History of Europe*. Cassell Paperbacks, Cassell & Co.

[Dierkes, M. *et al*, 2001] *Handbook of Organizational Learning and Knowledge*. Oxford University Press.

[Digital Planet, 2002] *The Global Information Technology*. World Information Technology and Services Allience.

[Dosi, G., Giannetti, R. and Toninelli, P. 1992] *Technology and Enterprise in a Historical Perspective*. Oxford Univ. Press.

[Dubois, D. and Prade, H. 1985] A review of fuzzy set aggregation connectives. *Information Sciences*, 36: 85-121.

[Dunnette, M.D. and Hough, L.M. 1990] *Handbook of Industrial and Organizational Psychology - vol.1*. Consulting Psychologists Press, Inc.

[Dunnette, M.D. and Hough, L.M. 1991] *Handbook of Industrial and Organizational Psychology - vol.2*. Consulting Psychologists Press, Inc.

[Dunnette, M.D. and Hough, L.M. 1992] *Handbook of Industrial and Organizational Psychology - vol.3*. Consulting Psychologists Press, Inc.

E

[Easterlin, R.A. 2000] The Worldwide Standard of Living Since 1800. *The Journal of Economic Perspectives*, 14 (1): 7-26.

[Etzioni, A. 1969] *A Sociological Reader on Complex Organizations*. Holt, Rinehrt and Winston, Inc.

F

[Fineman, S. 1993] *Emotions in Organizations*. SAGE Publications.

[Fogel, D.B. 2000] *Evolutionary computation: toward a new philosophy of machine intelligence*. IEEE Press.

[Forrester, J.W. 1961] *Industrial Dynamics*. The MIT Press.

[Forrester, J.W. 1973] *World Dynamics*. The MIT Press.

[Furukawa, K., Michie, D. and Muggleton, S. 1994] *Machine Intelligence: Machine Intelligence and Inductive Learning (13)*. Oxford University Press.

G

[Galbraith, J.R. 1973] *Designing Complex Organizations*. Addison-Wesley.

[Galbraith, J.R. 1977] *Organization Design*. Addison-Wesley.

[Galbraith, J.R. 2002] *Designing Organizations - An executive guide to strategy, structure, and process*. Jossey-Bass.

[Gaylord, R. and D'Andria, L. 1998] *Simulating Society: A Mathematica Toolkit for Modeling Socioeconomic Behavior*. Springer-Verlag.

[George, C.S. Jr. 1972] *The History of Management Thought*. Prentice-Hall.

[Gibbons, R. 1998] Incentives in Organizations. *The Journal of Economic Perspectives*, 12: 115-132.

[Gilbert, N. and Troitzsch, K.G. 1999] *Simulation for the Social Scientist*. Open University Press.

[Goodman, P.S. 1990] *Technology and Organizations*. Jossey-Bass Inc.

[Goleman, D. 1994] *Emotional Intelligence: Why it can matter more than IQ*. Bantam Books.

[Gordon, R.J. 2000] Does the New Economy Measure up to the Great Inventions of the Past? *The Journal of Economic Perspectives*, 14 (4): 49-74.

[Grinker, R.R. 1956] *Towards a Unified Theory of Human Behavior: An Introduction to General Systems Theory*. Basic Books Inc.

[Grusky, O. and Miller, G. 1981] *The Sociology of Organizations: Basic Studies*. The Free Press.

[Gul, F. 1997] A Nobel Prize for Game Theorists: The Contributions of Harsanyi, Nash and Selten. *The Journal of Economic Perspective*, 11 (3): 159-174.

[Gullahorn, J.T. and Gullahorn, J.E. 1963] A Computer Model of Elementary Social Behavior. *Behavioral Science*, 8 (1): 354-362.

[Gupta, M. and Sanchez, E. 1982] *Approximate Reasoning in Decision Analysis*. North-Holland.

H

[Haberstroh, C. 1965] Organization Design and Systems Analysis. In *Handbook of Organizations*, ed. by March, J. 1965: 1171-1211. Rand McNally & Company.

[Hagen, E.E. 1961] Analytical Models in the Study of Social Systems. *The American Journal of Sociology*, LXVII: 144-151.

[Haikonen, P.O. 2003] *The Cognitive Approach to Conscious Machines*. Imprint Academic.

[Hall, A.D. and Fagen, R.E. 1956] Definition of System. In *Modern Systems Research for the Behavioral Scientist*, ed. by Buckley, W. 1968: 81-92. Aldine Publishing Company.

[Halmos, P.R. 1960] Naive Set Theory. D. Van Nostrand. New Jersey.

[Hellriegel, D., Slocum, J.W. and Woodman, R.W. 2001] *Organizational Behavior*. South-Wester College Publishing.

[Helm, C. 2000] *Economic Theories of International Environmental Cooperation: New Horizons in Environmental Economics*. Edward. Elgar.

[Herbsleb, J., *et al*, 1994] Benefits of CMM-Based Software Process Improvement: Executive Summary of Initial Results. *Special Report CMU/SEI-94-SR-013*. September 1994.

[Hertz, J., Palmer, R. and Krogh, A. 1991] *Introduction to the Theory of Neural Computation*. Westview Press.

[Heyes, C. and Huber, L. 2000] *The Evolution of Cognition*. The MIT Press.

[Hodge, B.J., Anthony, W.P. and Gales, L.M. 2003] *Organization Theory - A Strategic Approach*. Prentice-Hall.

[Huxley, J., Hardy, A., and Ford, E. 1958] *Evolution as a Process*. George Allen & Unwin Ltd.

[Huxley, J. 1974] *Evolution: The Modern Synthesis*. George Allen & Unwin Ltd.

J

[Jager, R. 1995] Fuzzy Logic in Control. *PhD Thesis*, Delft University of Technology, Electrical Engineering Dept., Delft, The Netherlands.

[Johnson, D.G. 2000] Population, Food, and Knowledge. *The American Economic Review*, 90 (1): 1-14.

[Johnson, S. 2001] Emergence - *The connected lives of ants, brains, cities and software*. Allen Lane, The Penguin Press.

[Jones, C.I. 1997] On The Evolution of the World Income Distribution. *The Journal of Economic Perspectives*, 11 (3): 19-36.

K

[Karwowski, W. and Mital, A. 1986] *Applications of Fuzzy Set Theory in Human Factors. Advances in Human Factors/Ergonomics*, vol.6. Elsevier Science Publishers.

[Keltner, D. and Gross, J. 1999] Functional Accounts of Emotions. *Cognition and Emotion*, 13 (5):

467-480.

[Keltner, D. and Haidt, H. 1999] Social Functions of Emotions at Four Levels of Analysis. *Cognition and Emotion*, 13 (5): 505-521.

[Khandwalla, P.N. 1977] *Design of Organizations*. Harcourt Brace Jovanovich.

[Kipnis, D. 1990] *Technology and Power*. Springer-Verlag New York Inc.

[Klahr, D. and Kotovsky, K. 1989] *Complex Information Processing - The Impact of Herbert A. Simon*. Lawrence Erlbaum Associates, Inc.

[Klir, G.J. 1969] *An Approach To General Systems Theory*. Van Nostrand Reinhold Company.

[Klir, G.J. 1972] *Trends in General Systems Theory*. John Wiley & Sons.

[Klir, G.J. and Folger, T.A. 1992] *Fuzzy Sets, Uncertainty, and Information*. Prentice-Hall.

[Kornai, J. 2000] What the Change of System from Socialism to Capitalism Does and Does Not Mean. *The Journal of Economic Perspectives*, 14 (1): 27-42.

[Koza, J.R. 1992] *Genetic Programming: On the programming of computers by means of natural selection*. The MIT Press.

[Krajewski, L.J. and Ritzman, L.P. 2001] *Operations Management: Strategy and Analysis*. Addison-Wesley.

L

[La Porte, T.R. 1975] *Organized Social Complexity: Challenges To Politics and Policy*. Princeton University Press.

[Langton, 1995] *Artificial Life: An Overview*. The MIT Press.

[Lazear, E.P. 1998] *Personnel Economics for Managers*. John Wiley & Sons, Inc.

[Lee, C.C. 1990] Fuzzy Logic Control Systems: Fuzzy Logic Controllers – Part I and II. IEEE *Trans. on Systems, Man and Cybernetics*, vol.20 (2): 404-435.

[Lefrançoies, G. 1995] *Theories of Human Learning*. Brooks Cole Publishing Company.

[Leavitt, H. 1965] Applied Organizational Change in Industry: Structural, Technological and Humanistic Approaches. In *Handbook of Organizations*, ed. by March, J. 1965: 1144-1170. Rand McNally & Company.

[Lucking-Reiley, D. and Spulber, D.F. 2001] Business-to-Business Electronic Commerce. *The Journal of Economic Perspectives*, 15 (1): 55-68.

[Luger, G.F. and Stubblefield, W.A. 1998] *Artificial Intelligence: Structures and Strategies for Complex Problem Solving*. The Benjamin/Cummings Publishing Company, Inc.

M

[Malone, T.W. 1986] Modeling Coordination in Organizations and Markets. In *Readings in Distributed Artificial Intelligence*, by Bond and Gasser, 1988: 151-158. Morgan Kaufmann Publishers, Inc.

[Mamdani, E.H. 1974] Application of fuzzy algorithms for control of simple dynamic plan. *Proceedings of the IEE*, vol.121 (12): 1585-1588.

[March, J.G. 1965] *Handbook of Organizations*. Rand McNally & Company.

[March, J.G. 1994] *A Primer on Decision Making: How Decisions Happen*. The Free Press.

[March, J.G. 1998] *Decisions and Organizations*. Basil Blackwell.

[March, J.G. 1999] *The Pursuit of Organizational Intelligence*. Blackwell Business.

[March, J.G. and Olsen, J.P. 1975] The Uncertainty of the Past: Organizational Learning under

189

Ambiguity. *European Journal of Political Research*, (3): 147-171.

[March, J.G. and Simon, H.A. 1958] *Organizations*. 1st Ed. John Wiley & Sons, Inc.

[March, J.G. and Simon, H.A. 1993] *Organizations*. 2nd Ed. John Wiley & Sons, Inc.

[Markley, O.W. 1967] A Simulation of the SIVA Model of Organizational Behavior. *The American Journal of Sociology*, 73: 339-347.

[McKinlay, J.B. 1975] *Processing People: cases in organizational behaviour*. Holt-Blond Ltd.

[Michon, J.A. and Akyürek, A. 1992] *SOAR: A Cognitive Architecture in Perspective*. Kluwer Academic Publishers.

[Milgrom, P. and Roberts, J. 1992] *Economics, Organizations & Management*. Prentice-Hall Inc.

[Miller, J.G. and Miller, J.L. 1990] Introduction: The Nature of Living Systems. *Behavioral Science*, 35 (3): 157-163.

[Minsky, M. 1986] *The Society of Mind*. Picador.

[Mitchell, T.M. 1997] *Machine Learning*. The McGraw-Hill Companies, Inc.

N

[Neummann, J. von and Morgenstern, O. 1944] *Theory of Games and Economic Behavior*. Princeton University Press.

[Newell, A. and Simon, H.A. 1972] *Human Problem Solving*. Prentice-Hall.

[Newell, A. 1990] *Unified Theories of Cognition*. Harvard University.

[Nobre, F.S. 1997] Design and Analysis of Fuzzy Logic Controllers. *M.Sc. Thesis Dissertation*, 110 pages. Faculty of Electrical and Computer Engineering / State University of Campinas (UNICAMP), Brazil.

[Nobre, F.S. and Palhares, A.G. 1997] Qualitative and Quantitative Information in the Analysis and Design of Fuzzy Logic Controllers. *The Brazilian Society Journal of Automatic Control* (SBA), vol.8, no.2, pp.77-93.

[Nobre, F.S. and Nakasone, J. 1999] A Fuzzy Computational Approach for Evaluating Process Control Quality. *IEEE Proceedings of the International Conference on Fuzzy Systems*: 1701-1706. Seoul, Korea.

[Nobre, F.S. and Volpe, R. 1999] SEI-CMM Implementation at the NEC Brasil S.A. *Proceedings of the International Conference on Software Technology: Industrial Track 45-72*. Curitiba, Brazil.

[Nobre, F.S., Volpe, R. et al 2000] The Role of Software Process Improvement into TQM: An Industrial Experience. *IEEE Proceedings of the International Engineering Management Conference: 29-34*. Albuquerque-NM, USA.

[Nobre, F.S. *et al* 2000] Fuzzy Logic in Management Control: A Case Study. *IEEE Proceedings of the International Engineering Management Conference*: 414-419. Albuquerque-NM, USA.

[Nobre, F.S. and Steiner, S.J., 2001a] Fuzzy Logic in Organization Analysis and Control. *Proceedings of the International Conference in Fuzzy Logic and Technology: 126-129*. Leicester, England.

[Nobre, F.S. and Steiner, S.J. 2001b] Towards Intelligent and Immersive Manufacturing Systems. *Proceedings of the UK Workshop on Computational Intelligence: 232-236*. Edinburgh, UK.

[Nobre, F.S. and Steiner, S.J. 2002] Beyond the Thresholds of Manufacturing: Perspectives on Management, Technology and Organizations. *Proceedings of the IEEE International Engineering Management Conference: 788-793*. Cambridge-UK.

[Nobre, F.S. and Steiner, S.J., 2002] Fuzzy Logic in Control: A Tutorial. *Proceedings of the Postgraduate Research Symposium: 61-65*. ISBN: 0704423413. School of Engineering of The University of Birmingham, England.

190

[Nobre, F.S. and Steiner, S.J. 2003a] Perspectives on Organizational Systems: Towards a Unified Theory. *Doctoral Consortium on Cognitive Science at the ICCM 2003.* Bamberg-Germany, April 09th 2003.

[Nobre, F.S. and Steiner, S.J. 2003b] Beyond Bounded Rationality: Towards Economic Decision-Making Machines. *Conference on Dynamical Systems Approaches to Cognitive and Consciousness. Proceeding: 31.* Switzerland.

[Nobre, F.S. 2004] Analysis and Design of Organizational Systems: Towards a Unified Theory. *Oxford Centre for Brazilian Studies & 1st ABEP Conference. Proceeding: 52.* Oxford-UK.

[Nwana, H.S. and Azarmi, N. 1997] *Software Agents and Soft Computing: Towards Enhance Machine Intelligence.* Springer.

O

[Oliner, S.D. and Sichel, D.E. 2000] The Resurgence of Growth in the Late 1990s: Is Information Technology the Story? *The Journal of Economic Perspectives,* 14 (4): 3-22.

P

[Parsons, T. 1960] *Structure and Processes in Modern Societies.* Free Press.

[Paulk, M.C. *et al* 1994] *The Capability Maturity Model: Guidelines for Improving the Software Process.* Addison Wesley Longman, Inc.

[Paulk, M.C. and Chrissis, M.B. 2000] The November 1999 High Maturity Workshop. *Special Report CMU/SEI-2000-SR-003.* March 2000.

[Pedrycz, W. and Gomide, F. 1998] *An Introduction To Fuzzy Sets: Analysis and Design.* The MIT Press.

[Perrow, C. 1974] Zoo story or Life in the organizational sandpit. In *Perspectives on Organizations.* The Open University Press. ISBN 0335015689.

[Pine II, B.J. 1993] *Mass Customization: The New Frontier in Business Competition.* Harvard Business School Press.

[Plutchik, R. 1982] A psychoevolutionary theory of emotions. *Social Science Information,* 21: 529-553.

[Prietula, M.J., Carley, K. and Gasser, M. 1998] *Simulating Organizations: Computational Models of Institutions and Groups.* AAAI Press / The MIT Press.

[Pritchett, L. 1997] Divergence, Big Time. *The Journal of Economic Perspectives,* 11 (3): 3-17.

[Pugh, D.S. 1997] *Organization Theory: Selected Readings.* Penguin Books.

[Pugh, D.S. and Hickson, D.J. 1997] *Writers on Organizations.* Penguin Books.

R

[Rabin, M. 2002] A Perspective on Psychology and Economics. *European Economic Review,* 46: 657-685.

[Rapoport, A. 1986] *General Systems Theory: Essential Concepts & Applications.* Abacus Press.

[Reed, S.K. 1988] *Cognition: Theory and Applications.* 2nd Ed. Brooks-Cole Publishing Company.

[Reisberg, D. 1997] Cognition*: Exploring the Science of the Mind.* W.W. Norton & Company.

[Richter, M.N. 1982] *Technology and Social Complexity.* State University of New York.

[Rousseau, D.M. 1997] Organizational Behavior in the New Organizational Era. *Annual Review of Psychology,* 48: 515-546.

S

[Sanchez, E. and Zadeh, L.A. 1987] *Approximate Reasoning in Intelligent Systems, Decision and*

Control. Pergamon Press.

[Scarbrough, H. and Corbett, J.M. 1992] *Technology and Organization - Power, meaning and design*. Routledge.

[Scherer, K. R. 1982] Emotion as a process: Function, origin, and regulation. *Social Science Information*, 21: 555-570.

[Schmidt, F.L. and Hunter, J.E. 2000] Select Intelligence. In *The Blackwell Handbook of Principles of Organizational Behavior*. Ed. by Locke, E.A. 2000, p.3-14. Blackwell Publishers Ltd.

[SEI-CMU, 2004] *Process Maturity Profile: Software CMM*. August 2004. Software Engineering Institute of the Carnegie Mellon University (http://www.sei.cmu.edu).

[Shannon, C.E. and Weaver, W. 1963] *The Mathematical Theory of Communication*. University of Illinois Press.

[Scott, W.R. 1998] *Organizations: Rational, Natural, and Open Systems*. Prentice Hall, Inc.

[Scott, W.R. 2001] *Institutions and Organizations*. SAGE Publications.

[Shy, Oz 2001] *The Economics of Network Industries*. Cambridge University Press.

[Silvermann, D. 1970] *The Theory of Organizations*. Heinemann.

[Simon, H.A. 1952] A Formal Theory of Interaction in Social Sciences. *American Sociological Review*, 17: 202-211.

[Simon, H.A. 1957] *Models of Man: Social and Rational*. John Wiley & Sons.

[Simon, H.A. 1977] The *New Science of Management Decision*. Prentice-Hall, Inc.

[Simon, H.A. 1982a] *Models of Bounded Rationality: Economic Analysis and Public Policy. Vol.1*. The MIT Press.

[Simon, H.A. 1982b] *Models of Bounded Rationality: Behavioral Economics and Business Organization. Vol.2*. The MIT Press.

[Simon, H.A. 1983] *Reason in Human Affairs*. Stanford University Press.

[Simon, H.A. 1996] *The Sciences of the Artificial*. 3rd Ed. The MIT Press.

[Simon, H.A. 1997a] *Models of Bounded Rationality: Empirically Grounded Economic Reason. Vol.3*. The MIT Press.

[Simon, H.A. 1997b] *Administrative Behavior: A Study of Decision-Making Processes in Administrative Organizations*. The FREE Press.

[Simon, H.A. 2002] Organization theory in the age of computers and electronic communication networks. In *The Economics of Choice, Change and Organization - Essays in Memory of Richard M. Cyert*. Ed. by Augier, M. and March, J.M. 2002, p.404-418. Edward Elgar.

[Sims, D., Fineman, S. and Gabriel, Y. 1993] *Organizing & Organizations: An Introduction*. SAGE Publications.

[Singh, J.V. 1990] *Organizational Evolution: New Directions*. SAGE Publications.

[Stacey, R.D., Griffin, D. and Shaw, P. 2000] *Complexity and Management - Fad or radical challenge to systems thinking?* Routledge.

[Starbuck, W.H. 1965] Mathematics and Organization Theory. In *Handbook of Organizations*, ed. by March, J.G. 1965: 335-386. Rand McNally & Company.

T

[Taylor, F.W. 1911] *The Principles of Scientific Management*. New York: Harper.

[Tiwana, A. 2001] *The Essential Guide to Knowledge Management: E-Business and CRM Applications*. Prentice-Hall.

192

[Turing, A.M. 1950] Computing Machinery and Intelligence. *Mind - A Quarterly Review of Psychology and Philosophy*, LIX (236): 433-460.

V

[Vecchio, R.P. 1995] *Organizational Behavior*. Harcourt Brace & Company.

W

[Wang, L. 1994] *Adaptive Fuzzy Systems and Control: Design and Stability Analysis*. PTR Prentice-Hall.

[Watt, S.N.K. 1997] Artificial Societies and Psychological Agents. In *Software Agents and Soft Computing: Towards Enhancing Machine Intelligence*, by Nwana and Azarmi, 1997, p.27-41. Springer.

[Weiss, G. 1999] *Multiagent Systems – A Modern Approach to Distributed Artificial Intelligence*. The MIT Press.

[Wiener, N. 1948] *Cybernetics*. 1st Ed. The MIT Press.

[Wiener, N. 1954] *The human use of human beings: cybernetics and society*. 2nd Ed. London.

[Wiener, N. 1961] *Cybernetics or control and communication in the animal and the machine*. 2nd Ed. The MIT Press.

[Williamson, O.E. and Masten, S.E. 1999] *The Economics of Transaction Costs*. Edward Elgar Publishing, Inc.

[World Bank, 2003] *The Little Data Book*. World Bank. ISBN 0-8213-5426-4.

[Wren, D.A. 1987] *The Evolution of Management Thought*. 3rd Ed. John Wiley and Sons.

Z

[Zadeh, L.A. 1962] From Circuit Theory to System Theory. *Proceedings of the IRE*, 50: 856-865.

[Zadeh, L.A. 1965] Fuzzy Sets. *Information and Control*, 8: 338-353.

[Zadeh, L.A. 1968] Fuzzy Algorithms. *Information and Control*, 12: 94-102.

[Zadeh, L.A. and Polak, E. 1969] *System Theory*. McGraw-Hill.

[Zadeh, L.A. 1972] A Rationale for Fuzzy Control. *Transactions of the ASME: Journal of Dynamic Systems, Measurements, and Control*, March: 3-4.

[Zadeh, L.A. 1973] Outline of a New Approach to the Analysis of Complex Systems and Decision Process. *IEEE Transactions on Systems, Man, and Cybernetics*, 3 (1): 28-44.

[Zadeh, L.A. 1975] The concept of a linguistic variable and its application to approximate reasoning: part I and II. *Fuzzy Sets and their Applications: Selected Papers by L.A. Zadeh*. Ed. by Yager, R.R. et al. (1987): 219-327. John & Sons.

[Zadeh, L.A. 1976] The concept of a linguistic variable and its application to approximate reasoning: part III. *Fuzzy Sets and their Applications: Selected Papers by L.A. Zadeh*. Ed. by Yager, R.R. et al. (1987): 329-366. John & Sons.

[Zadeh, L.A. 1988] Fuzzy Logic. *IEEE Computer*, April: 83-92.

[Zadeh, L.A. 1994] Soft Computing and Fuzzy Logic. *IEEE Software*, November: 48-56.

[Zadeh, L.A. 1996a] Fuzzy Logic = Computing with Words. *IEEE Transactions on Fuzzy Systems*, 4 (2): 103-111.

[Zadeh, L.A. 1996b] The Evolution of Systems Analysis and Control: A Personal Perspective. *IEEE Control Systems*, June: 95-98.

[Zadeh, L.A. 1997] The Roles of Fuzzy Logic and Soft Computing in the Conception, Design and Development of Intelligent Systems. In *Software Agents and Soft Computing: Towards Enhancing*

Machine Intelligence, ed. by Nwana and Azarmi 1997: 183-190. Springer.

[Zadeh, L.A. 1999] From Computing with Numbers to Computing with Words – From Manipulation of Measurements to Manipulation of Perceptions. *IEEE Transactions on Circuits and Systems*, 45 (1): 105-119.

[Zadeh, L.A. 2001] A New Direction in AI: Toward a Computational Theory of Perceptions. *AI Magazine*. Spring: 73-84.